Fertiliser Manual (RB209)

8th Edition

June 2010

Published by TSO (The Stationery Office), part of Williams Lea,
and available from:

Online
www.tsoshop.co.uk

Mail, Telephone, Fax & E-mail
TSO
PO Box 29, Norwich, NR3 1GN
Telephone orders/General enquiries: 0333 202 5070
Fax orders: 0333 202 5080
E-mail: customer.services@tso.co.uk
Textphone 0333 202 5077

TSO@Blackwell and other Accredited Agents

Published with the permission of the Department for Environment, Food and Rural Affairs on behalf of the Controller of Her Majesty's Stationery Office

© Crown Copyright 2010

Copyright in the typographical arrangement and design is vested in the Crown. Applications for reproduction should be made in writing to the Office of Public Sector Information, Information Policy Team, Kew, Richmond, Surrey, TW9 4DU.

First published 2010
Tenth impression 2016

ISBN 978 0 11 243286 9

Text and cover printed on Cocoon 50% Recycled Silk, a recycled paper manufactured from 50% recycled fibre and 50% virgin fibre at a mill certified with ISO14001 environmental management standard.

The pulp is bleached using an elemental chlorine free (ECF) process.

Printed in the United Kingdom for The Stationery Office.
P002800865 C2.5 04/16

PB13381

June 2010

Fertiliser Manual (RB209) 8th Edition errata

Section 1: Principles of nutrient management and fertiliser use
- Page 40 (Section 1): In the second table, the value for 8t/ha at Index 0 should be 155, not 185.

Section 2: Organic Manures
- Top of pages 67 and 69: The top line of text on each page should be on the bottom of pages 66 and 68, respectively, i.e. they are a continuation of the footnotes.
- Page 67: In the second table, Pig slurry – Total and available nutrients, the 6% dry matter, 4.4 kg/t total N and 2.8 kg/t readily available N numbers should not be in bold.
- Page 72: The '2.6', '1.2', '3.2' numbers in row one should be in row two. Also, the '0.9[a]' nitrogen number in the next line of numbers should be one row up, i.e. in the row for 'Nitrogen (table on page 66)'.
- Page 73: The '0.7[a]' nitrogen number should be one row up.
- Page 78: Under the nitrogen column the '1.7[a]' number should be one row up.
- Page 81: Under Stage 6 the 'NIL', '20' and '70' numbers should be one row down i.e. on the Stage 3 minus Stage 4 line. The '£54/ha' should also move down a row.

Section 3: Using the recommendation tables
- Page 94: Table D, header D2. 'Other mineral soils' should read 'Medium soils'.
- Page 98: In the Example 2 box, the pig slurry supplies 65 kg/ha of crop available nitrogen not 86. Thus 120 – 65 = 55 kg N/ha as fertiliser should be applied.

Section 4: Arable and forage crops
- Pages 106, 116: The footnote to the tables should refer readers to page 48, not page 30.
- Pages 113 – All cereals, 118 – Oilseed rape, 123 – Potatoes, 127 – Forage maize, 129 – Other forage crops: The sentence "No adjustment for yield should be made where the soil Index is higher than target" should be deleted.
- Page 114: The final line, column 4, of the table should read '105 (2-) 75 (2+)' not 105 (2-) 75 (2-).
- Page 114 (Sulphur): Range of sulphate-containing fertiliser applied in early spring should be '25-50 kg SO_3/ha' not 25-40.
- Page 115 (The effect of economic changes on nitrogen rates): The figures in the table on p. 116 have been revised – see page overleaf.
- Pages 120, 123, 126, 127, 129, 131: The tables on these pages should show a heading under the Index values 'kg/ha' not 'kg N/ha'.

March 2011

Fertiliser Manual (RB209) 8th Edition errata

Section 8: Grass

- General point: In the Grass section of the Fertiliser Manual, the nitrogen recommendation tables allow for concentrate use. If bulky, relatively moist, bought-in feeds are used on the farm, these can be treated as concentrates by entering their use on a dry-matter basis. For example, feeding 2.5 t/cow brewers grains at 23% DM would be equivalent to 0.6 t/cow concentrates.
- Page 188: The list of soil types for Medium Soil Available Water should include 'organic' soils
- Page 189: In the Calculating stocking rate paragraph, the last sentence should read:

  "Calculate the average number of livestock as LU and divide by the total area of grass and forage crops in the enterprise to give stocking rate."

  Total area of grass and forage crops (including forage maize) should be used, not just the area of grassland.
- Pages 200-201: Tables 8.4 and 8.5 have been revised – see page overleaf.

Appendix 2

- Pages 221-224: In the title for Appendix 2, 'Smn' should be 'SMN'.

Appendix 7

- Page 234: The 'SO_3' under 'Kainit' and the 'Na_2O' under 'Sylvinite' should be indented so they appear in the right hand column.

Appendices 9 and 10

- General point: Readers should note that these two Appendices are interlinked. Values for the variables and parameters in the equations in Appendix 9 are to be found in the table in Appendix 10.
- Page 239: The '150.0' t/ha for carrots in column 2 is based on yield of ungraded roots removed from field, not fresh market yield.

March 2011

Fertiliser Manual (RB209) 8th Edition errata

Table page 116

	Fertiliser N content (%)	Fertiliser Cost (£/tonne Product)					
Ammonium Nitrate	34.5%	£138	£207	£276	£345	£414	£483
Urea[a]	46.0%	£184	£276	£368	£460	£552	£644
Urea-Ammonium Nitrate Liquid[a]	28.0%	£112	£168	£224	£280	£336	£392
Cost of Fertiliser Nitrogen	£/kg N	£0.40	£0.60	£0.80	£1.00	£1.20	£1.40
Change to recommended N for Oilseed Rape (kg N/ha)							
Grain sale price (£/tonne)	200	20	-20	-50	-70	-80	-100
	225	30	-10	-30	-60	-70	-90
	250	40	0	-20	-50	-60	-80
	275	50	10	-10	-40	-50	-70
	300	60	20	-10	-30	-50	-60
	325	70	30	0	-20	-40	-50
	350	70	40	10	-10	-30	-50
	375	80	40	20	-10	-20	-40
	400	90	50	20	0	-20	-30
	425	90	50	30	10	-10	-30
	450	100	60	30	10	-10	-20

a. See notes on efficiency of use of different kinds of applied nitrogen page 48.

March 2011

Fertiliser Manual (RB209) 8th Edition errata

Table 8.4 Beef: Grass Growth Class *Average*. Whole-season total nitrogen requirement for cut and grazed grass (kg N/ha)

BEEF Grass Growth Class *Average*			Total N requirement		
	Concentrate use	Stocking Rate	Cut		Grazed
	(t/animal/year)	LU/ha	kg/ha	Indicative yield* (t DM/ha)	kg/ha
Intensively grazed (lowland dairy steers and heifers; some suckler herds)	0.4	1.8	320	9.6	300
		1.6	310	9.4	210
		1.4	290	9.2	150
Moderately grazed (upland and lowland suckler herds; lowland dairy steers and heifers)	0.2	1.4	310	9.4	170
		1.2	290	9.2	110
		1.0	280	9.0	50
Extensively grazed (moorland/hill beef; grazing for biodiversity)	0.0	0.9	240	8.3	0
		0.5	210	7.9	0
		0.3	200	7.6	0

Table 8.5 Beef: Grass Growth Class *Very poor/Poor*. Whole-season total nitrogen requirement for cut and grazed grass (kg N/ha)

BEEF Growth Class *Very poor/Poor*			Total N requirement		
	Concentrate use	Stocking rate	Cut		Grazed
	(t/animal/year)	LU/ha	kg/ha	Indicative yield* (t DM/ha)	kg/ha
Intensively grazed (lowland dairy steers and heifers; some suckler herds)	0.4	1.3	270	6.8	270
		1.2	260	6.7	220
		1.1	250	6.7	180
Moderately grazed (upland and lowland suckler herds; lowland dairy steers and heifers)	0.2	1.2	270	6.9	250
		1.1	270	6.8	200
		1.0	260	6.7	160
Extensively grazed (moorland/hill beef; grazing for biodiversity)	0.0	0.9	230	6.3	110
		0.5	210	5.9	0
		0.3	200	5.8	0

March 2011

Foreword

Farming is an important feature of our way of life. It shapes the landscape we enjoy, provides much of our food and is a vital part of maintaining and improving a healthy, thriving and diverse natural environment.

The agriculture sector faces a number of challenges. Demand for crops will increase as the global population rises. However, agricultural practices including the use of fertilisers have the potential to damage the natural environment by polluting water courses, releasing greenhouse gases and ammonia to the air and by damaging the soil.

Using this manual will help farmers and land managers better assess the fertiliser required for the range of crops they plan to grow, by **suggesting what level of nutrients are required to provide the best financial return for the farm business**. The manual will help ensure that proper account is taken of both mineral fertilisers and other sources of nutrients such as manures and slurries, so helping to prevent costly over-application.

Deciding the correct amount of nutrient to use along with planning how they are managed and applied, supported where appropriate by specialist advice, are important factors in good nutrient management. This manual sits alongside other government advice and guidance such as the Code of Good Agricultural Practice ("CoGAP") and guidance from the agricultural sector such as *"Tried and Tested"*.

Good planning and the right choice of farming practice can help minimise any impacts on the environment, help maintain and improve crop yield and also save farmers and land managers money.

Acknowledgements

The revision of the former *Fertiliser Recommendations for Agricultural and Horticultural Crops* (RB209) to create this *Fertiliser Manual* has been led by Rothamsted Research, North Wyke Research and Warwick-HRI, compiling information from many sources, and with guidance from a Steering Group comprising representatives of the main user organisations. Defra gratefully acknowledges the support and generous contributions of the following organisations for their work on the Steering Group.

ADAS
Agri-Food and Biosciences Institute for Northern Ireland (AFBINI)
Agriculture and Horticulture Research Forum – represented by HGCA
Agricultural Industries Confederation (AIC)
Association of Independent Crop Consultants (AICC)
British Institute of Agricultural Consultants (BIAC)
Country Land and Business Association (CLA)
Environment Agency (EA)
Fertiliser Advisers Certification and Training Scheme (FACTS)
National Assembly for Wales Agriculture Department (NAWAD)
National Farmers Union (NFU)
North Wyke Research (NWRes)
Potash Development Association (PDA)
Rothamsted Research (RRes)
Scottish Agricultural College (SAC)
Scottish Government

During the revision, a wide range of research information was considered from projects funded mainly by Defra, the Levy bodies, Water UK and the fertiliser industry. The key research organisations involved in the technical working groups with Rothamsted Research, North Wyke Research and Warwick-HRI were ADAS, the Association of Independent Crop Consultants (AICC), East Malling Research, Ecopt and The Arable Group (TAG).

The contributions of the many organisations and individuals involved in the revision process is gratefully acknowledged. The Steering Group also gratefully acknowledges the contribution of the reviewers of the main sections; Peter Gregory, Bryan Davies, Philip White, Alan Brewer, George Fisher. Panel members advised on the sections relating to their expertise and practical experience but the panel did not see the final draft of the Fertiliser Manual before its release. Additionally, the following organisations made a significant contribution to one or more sections of this *Fertiliser Manual*:

Enviros
Imperial College
Association for Organic Recycling
WRAP

The Basis of Good Practice

Reliable information
- Soil type
- Field cropping, fertilising and manuring history
- Regular soil analysis for pH, P, K and Mg
- Nutrient balances - surplus or deficit from applications to previous crops
- Soil analysis for soil mineral nitrogen
- Winter rainfall
- Crop tissue analysis where appropriate (e.g. for fruit crops and for sulphur and potassium on grassland)

Correct decisions on on-farm economic optimum crop yields and the requirement for applied nutrients
- Take account of fertiliser nitrogen and crop produce prices
- Consider market requirements for quality and quantity of harvested produce
- Consider adjusting phosphate and potash for yield level

Assessment of available nutrients from organic manures
- Apply manures in spring if possible, and incorporate rapidly into the soil following broadcast application to tillage land
- Make use of manure analysis (on-farm and laboratory testing)
- Calculate available nutrients based on manure type, method and time of application

Decisions on the rate, method and timing of fertiliser application for individual crops
- Apply nitrogen to meet periods of greatest demand for nitrogen
- Consider placement of fertilisers for responsive crops

Careful selection of fertilisers
- Consider the cost effectiveness of alternative fertiliser materials
- Take account of the nutrient percentage and the availability of nutrients for crop uptake
- Make sure that the physical quality of the fertiliser will allow accurate spreading

Accurate application of fertilisers and manures
- Regularly maintain and calibrate fertiliser spreaders and sprayers
- Regularly check and maintain manure spreaders

Record keeping
- Keep accurate field records to help with decisions on fertiliser use

This *Manual* gives information for all of the above stages, including a list of sources of other useful information.

Summary of Main Changes from Previous Edition

Overall presentation

- To make its use as easy as possible, the layout of the *Manual*, Section and Appendix numbers remain largely the same as those in the 7th Edition of RB209: Section 7 covers Biomass Crops and Section 8 covers Grass. However, every section has been fully revised, making use of all available peer-reviewed data and with full and comprehensive consultation.

General information

- Section 1, describing the principles of nutrient management and fertiliser use, has been extensively revised. More information on meeting environmental objectives is included.

Soil Nitrogen Supply (SNS) Index system

- The SNS system has been fully revised. Definitions of soil type have been made clearer and users are directed to a flowchart and a table to make identification easier.
- The SNS Index tables have been revised. Some crops, such as 'Uncropped land' (previously 'Set-aside') have moved down an Index.

Revised nitrogen recommendations

- Optimum nitrogen requirements for all arable crops have been revised and those for wheat are now based on a breakeven ratio of 5:1 (cost of nitrogen as p/kg N divided by value of crop as p/kg yield or, put another way, 5 kg of grain are needed to pay for 1 kg N.).

Updated organic manures section

- The Organic Manures section has been expanded to include nutrient contents for more types of livestock manures, biosolids, composts and some other organic materials that are commonly applied to agricultural land ('content' is used rather than the more accurate 'concentration' as this is common usage).
- The nutrient contents of livestock manures and biosolids have been updated and expanded guidance provided on their readily available nitrogen contents.
- MANNER-*NPK* has been used to provide updated guidance on crop available nitrogen.
- The table of excreta and nutrients produced during the period when livestock are housed has been updated.
- Enhanced guidance has been provided on the benefits of using bandspreading and shallow injection slurry application methods.
- Advice on how to interpret laboratory analyses has been provided and manure sampling guidelines updated.

Summary of Main Changes from Previous Edition

New approach to grassland recommendations

- For grassland the amounts are based on a new *systems based approach* which helps determine the levels of fertiliser required to produce the grass needed in different farming systems at varying levels of intensity. This approach recognises that for many situations fertiliser application rates for grassland need not be based solely on the on-farm economic optimum. The grassland chapter in this Manual lists the main differences between the new approach and the grassland chapter in the 7th edition of RB209.

New recommendations for biomass crops

- Recommendations for biomass crops are included for the first time.

Potatoes

- Average nitrogen recommendations have decreased slightly. Phosphate and magnesium recommendations have been reduced, but there are slightly increased recommendations for potash.

Sugar Beet

- Nitrogen recommendations have been increased slightly, particularly for an SNS Index of greater than 2. It has been made clear that, if growers want to achieve the optimum yield they will need to ensure that the P and K Index is at least 2.

Sulphur recommendations

- An up-to-date map of sulphur deposition has been supplied by the NERC Centre for Ecology and Hydrology.

- The risk of sulphur deficiency is becoming more widespread. Recommendations for the diagnosis and prevention of deficiency remain the same but the need for sulphur fertilisers is emphasised.

Phosphate and potash recommendations

- Emphasis is placed on maintaining soil target Indices and replacing offtake in harvested crops.

- Build-up applications at Indices 0 and 1 have been increased slightly and run-down applications at Indices above target have been decreased.

Contents

	Page
Foreword	1
Acknowledgements	2
The Basis of Good Practice	3
Summary of Changes	4
Introduction	10
Section 1: Principles of nutrient management and fertiliser use	15
Crop nutrient requirements	16
Integrated plant nutrient management	17
Important soil properties	17
Soil acidity and liming	19
Nitrogen (N) for field crops	22
Phosphate, potash and magnesium for field crops	33
Sulphur	43
Sodium	44
Micronutrients (trace elements)	45
Fertiliser types and quality	46
Fertiliser application	49
Protection of the environment	50
Section 2: Organic Manures	55
Introduction	56
Livestock manures	56
Allowing for the nutrient content of livestock manures	61
Cattle, pig, sheep, duck or horse solid manures – total and available nutrients	62
Poultry manures – total and available nutrients	64
Cattle slurry and dirty water – total and available nutrients	65
Pig slurry – total and available nutrients	67
Using livestock manures and fertilisers together	69
Sewage sludges (biosolids)	74
Biosolids – total and available nutrients	75
Allowing for the nutrient content of biosolids	77

Contents

Compost	79
Industrial wastes	82
Section 3: Using the recommendation tables	**85**
Finding the nitrogen recommendation	86
Field Assessment Method	86
Table A. Soil Nitrogen Supply (SNS) Indices for low rainfall areas (500-600 mm annual rainfall, 50-150 mm excess winter rainfall) – based on the last crop grown	91
Table B. Soil Nitrogen Supply (SNS) Indices for moderate rainfall areas (600-700 mm annual rainfall, or 150-250 mm excess winter rainfall) – based on the last crop grown	92
Table C. Soil Nitrogen Supply (SNS) Indices for high rainfall areas (over 700 mm annualrainfall, or over 250 mm excess winter rainfall) – based on the last crop grown	93
Tables D. Soil Nitrogen Supply (SNS) Indices following ploughing out of grass leys	94
Meaurement Method	95
Finding the phosphate, potash and magnesium recommendations	99
Finding the sulphur and sodium recommendation	101
Selecting the most appropriate fertiliser	102
Calculating the amount of fertiliser to apply	102
Section 4: Arable and Forage Crops	**103**
Checklist for decision making	104
Wheat, autumn and early winter sown – nitrogen	105
Barley, winter sown – nitrogen	107
Oats, rye and triticale, winter sown – nitrogen	109
Wheat, spring sown – nitrogen	110
Barley, spring sown – nitrogen	111
Oats, rye and triticale, spring sown - nitrogen	112
All cereals – phosphate, potash, magnesium and sulphur	113
Oilseed rape, autumn and winter sown – nitrogen	115
Oilseed rape and linseed, spring sown – nitrogen	117
Oilseed rape and linseed – phosphate, potash, magnesium and sulphur	118
Peas (dried and vining) and beans	120
Potatoes – nitrogen	121

Contents

Potatoes– phosphate, potash and magnesium	123
Sugar beet	125
Forage maize	127
Other forage crops	129
Ryegrass grown for seed	131
Section 5: Vegetables and Bulbs	**133**
Checklist for decision making	134
Fertiliser use for vegetables	135
Asparagus	138
Brussels sprouts and cabbage	139
Cauliflowers and calabrese	141
Celery and self blanching celery	143
Peas (market pick) and beans	144
Lettuce, radish, sweetcorn and courgettes	145
Onions and leeks	147
Root vegetables	149
Bulbs and bulb flowers	151
Section 6: Fruit, Vines and Hops	**153**
Checklist for decision making	154
Fertiliser use for fruit, vines and hops	155
Fruit, vines and hops, before planting	157
Top fruit, established orchards	159
Soft fruit and vines – established plantations	162
Leaf analysis for top and soft fruit	166
Apple fruit analysis	168
Hops	171
Section 7. Biomass Crops	**173**
Section 8. Grass	**177**
Differences between 7th and 8th editions of grasslands chapters of (RB209)	178
Check list for decision making	179
Principles of fertilising grassland	181
Protecting the environment	185

Contents

Finding the nitrogen recommendation	187
Finding the phosphate, potash and magnesium recommendation	190
Grass establishment – nitrogen	191
Grass establishment – phosphate, potash and magnesium	191
Cutting and Grazing for dairy production – nitrogen	192
Cutting and Grazing for beef production – nitrogen	198
Cutting and Grazing for sheep production – nitrogen	203
Grazing of grass/clover swards – nitrogen	208
Grazed grass – phosphate, potash and magnesium	210
Cutting of grass/clover swards, red clover and lucerne – nitrogen	210
Grass silage – phosphate, potash, magnesium and sulphur	211
Hay – nitrogen	213
Hay – phosphate, potash and magnesium	213

Section 9. Other Useful Sources of Information — 215

Appendix 1. Description of Soil Types — 219

Appendix 2. Sampling for Soil Mineral Nitrogen (SMN), Estimation of Crop Nitrogen Content and Soil Mineralisable Nitrogen — 221

Appendix 3. Sampling for Soil pH, P, K and Mg — 225

Appendix 4. Classification of Soil P, K and Mg Analysis Results into Indices — 227

Appendix 5. Phosphate and Potash in Crop Material — 228

Appendix 6. Sampling Organic Manures for Analysis — 230

Appendix 7. Analysis of Some Fertiliser and Liming Materials — 234

Appendix 8. Conversion Tables — 236

Appendix 9. Calculation of the Crop Nitrogen Requirement (CNR) for Field Vegetable Crops — 237

Appendix 10. Information for Derivation of Crop Nitrogen Requirement of Field Vegetable Crops — 239

Appendix 11. Profit maximisation and social costs — 240

Glossary — 243

Notes — 250

June 2010

Introduction

This *Manual* has been produced for use by agricultural consultants, farmers and their agents and all other individuals and organisations concerned with developing and maintaining a sustainable agricultural industry in England and Wales. The principles of crop nutrition and soil fertility are presented briefly as background to the recommendations for the use of fertilisers, organic manures, including biosolids and composts, and lime for field crops, grassland, vegetables, fruit and bioenergy crops. It is an essential source of reference and advice for farmers, FACTS Qualified Advisers and those seeking qualification.

Achieving on-farm optimum economic crop yields of marketable quality with minimum adverse environmental impact requires close attention to detail. The recommendations given in this *Manual* seek to do this by ensuring that the available supply of plant nutrients in soil is judiciously supplemented by additions of nutrients in fertilisers and those sources of organic amendments that are available on the farm. Crops must have an adequate supply of nutrients, and many crops show large and very profitable increases in yield from the correct use of fertilisers to supply nutrients, and liming to correct any soil acidity. The recommendations in this *Manual* support Defra's aims to ensure good agricultural practice for the supply of nutrients to support economically viable crop production in environmentally acceptable ways. It is particularly directed to the use of fertilisers and the recommended amounts are those that typically will give the best financial return for each given set of circumstances.

The fertiliser recommendations for grassland adopt a *new systems-based approach* which is based on the economic need to produce the amount of home grown forage necessary to maintain a target intensity of production, rather than the optimal amount relative to the cost of fertiliser. This enables farmers who may be operating at widely different stocking rates and feeding different levels of concentrates to obtain relevant recommendations for fertiliser nitrogen application.

This *Manual* is an authoritative source of advice in a book format that provides a national standard for England and Wales. Other systems, including computerised systems that are already available or being developed, may give equally good recommendations but the extra complexity to achieve a very similar recommendation may not be necessary.

The recommendations accord with Integrated Farm Management where the aim is to optimise the benefit from all inputs to achieve an efficient and profitable production system with minimum adverse environmental impact. Integrated nutrient management seeks to optimise the use of all available plant nutrients on the farm. This includes those in the soil and those that can be recycled by the return of organic manures, including sewage sludge (biosolids) and composts, and fertilisers that are used to supplement these sources. The recommendations in this *Manual* for the use of fertilisers are not appropriate for organic farming systems, where the use of soluble manufactured fertilisers is, with few exceptions, prohibited. Further information can be obtained from the appropriate organic sector body.

The principles of nutrient management and the basis of the recommendations are explained in Section 1. Section 2 gives detailed guidance on how to calculate the nutrients supplied from a specific manure application and how to maximise the value of the nutrients in a range of organic manures. Instructions on how to use the recommendation tables are in Section 3 and are based on the nutrient Index system. Sections 4 to 8 give detailed recommendations for all

Introduction

major field crops, grassland, vegetables, fruit/vines and hops and bioenergy crops. Section 9 gives references to other sources of authoritative and detailed information. Appendices cover a range of topics including a description of different soil types, sampling methods for soil and organic manures and the soil classification system currently in use.

Good Nutrient Management

Good nutrient management is essential to helping farmers grow the food we want to buy without harm to the environment or health: farming can produce food and other crops profitably, sustainably and with high environmental standards. Over the last two decades, farmers in the UK have succeeded in increasing yields while reducing use of fertiliser, and the associated greenhouse gas emissions. Better nutrient management has played a big part in achieving this. The standard of nutrient management needs to continue to improve if farming is to meet its future challenges: to become more competitive and profitable, produce more food but in a sustainable way that protects the environment and biodiversity and which preserves the natural resources on which future food production depends.

Maintaining a profitable farming business requires continued development and use of a wide range of skills by farmers and their advisers. **Good nutrient management** is an important aspect of this, and can contribute both to the efficiency of the farming business and to reducing environmental impacts. Defra already works with the farming and fertiliser sectors to support and encourage skills development. This new Fertiliser Manual provides a tool to help operate a profitable farm business whilst protecting the environment.

Research, development and innovation will help develop new techniques and technologies that along with enhancing skills will help farming adapt and tackle the challenges it faces. Sound evidence is so important and research establishments, industry and others will play an important role in helping the sector rise to the challenges it faces.

This manual will help meet the challenges of producing more of the food wanted by consumers without harming the environment. It contains guidance to help balance the benefits of fertiliser use against the costs – both economic and environmental. It explains why **good nutrient management** is about more than just the fertilisers you buy: it can save you money as well help protect the environment for future generations. To help your forward planning, it also outlines possible future changes that could further affect fertiliser use.

The growing challenges

The agricultural sector faces a number of challenges including producing more food to feed a growing population, whilst impacting less on the environment. The United Nations predicts that the world population will rise to more than nine billion by 2050.

Nutrient use on farms – whether from inorganic fertiliser, manures or digestate – can have a significant impact on the environment: locally, in terms of ground and surface water quality, and over a longer range and longer timescale by impacting on the quality of the air we breathe, affecting long-term soil quality and by contributing significantly to greenhouse gas emissions.

Introduction

Improving nutrient management will help improve both the profitability of farming and its net environmental impact. Ensuring the carefully managed application of all nutrients, including mineral fertilisers and organic materials such as slurries, manures, and digestate from anaerobic digestion systems, helps to close the 'nutrient gap' that arises as the crop uses nutrients to grow and is then removed at harvest. Optimising the production benefits by ensuring good nutrient uptake by crops helps minimise an excess in the soil where, for example, nitrogen can be lost as nitrous oxide or nitrate emissions.

Demand for land to grow non-food crops – such as biomass (for which recommendations are included in the Manual) – and from non-farming uses may well increase. This is one factor which may mean that higher crop yields will be necessary to supply the increased demand for food.

Low Carbon Farming

At present agriculture is estimated to contribute around 8% of total UK greenhouse gas emissions. An estimated 76% of the UK's nitrous oxide emissions – a greenhouse gas around 300 times more potent than carbon dioxide – come from agriculture. Soil nitrous oxide emissions come from three on-farm sources: soil microbial activity; organic manure applications and nitrogen fertiliser applications.

The amount of nitrous oxide released from spreading fertiliser varies according to the soil type, the weather conditions, when and how the spreading was done, and many other factors. Careful planning that maximises the efficiency of fertiliser use and better management of manures can help reduce how much nutrient – and hence money – is lost as nitrous oxide (or ammonia).

All industries have an important role in meeting the UK carbon budgets established by the UK Climate Change Act (2008), and the farming industry has published its Greenhouse Gas Action Plan, which confirms its intent to play its part in helping to reduce greenhouse gas emissions – through using fertilisers more efficiently, and by improving livestock feeding, breeding and manure management. This manual is one source of advice to help industry do this.

Water quality

Losses from the application of fertilisers and spreading of manures contribute to diffuse water pollution, due to the activities of many farms.

In England and Wales around 60% of nitrates and 25% of phosphates in our waters originate from agricultural land.[1] Elevated levels of these nutrients can harm the water environment and impact on biodiversity. In addition, excessive amounts of agricultural pollutants including nitrates and phosphates have to be removed before water can be supplied to consumers. The Water Framework Directive requires our rivers, lakes, ground and coastal waters to reach good ecological and chemical status by 2015. Farmers will need to manage their land carefully to play their part in achieving this.

---

1 Protecting our Water, Soil and Air: A Code of Good Agricultural Practice for farmers, growers and land managers, Defra 2009

Introduction

Soils

Policy development on soils aims to improve the sustainable management of soil and successfully tackle degradation of our soils.

To help protect our soils we want to reduce the levels of pollutants entering them from materials spread to land. Spreading organic (e.g. composts, manure, sewage sludge) and inorganic materials (e.g. recycled gypsum from waste plasterboard) materials to land plays an important role in increasing levels of organic matter in soil, reducing fertiliser requirements and diverting materials from landfill. It can also have important agricultural and ecological benefits as well as getting the best value from resources we rely on.

Sometimes materials spread on land can also contain low concentrations of pollutants, especially heavy metals, which following repeated applications can accumulate in the soil. This could pose a risk to human health and the environment. Remediating soils which contain pollutants is difficult and costly and so it is important to prevent unacceptable levels of pollutants getting in to the soil in the first place.

Air quality

Agricultural activities account for around 90% of the ammonia emissions to air. Ammonia (a compound of nitrogen and hydrogen) plays a complex role in a number of different air quality issues. High concentrations of ammonia in the air can mean that nitrogen is deposited from the air onto the land, and this can damage some habitats, by changing the species of plants present and also damaging streams, rivers and other water bodies. The ammonia can also be re-released from soils as nitrous oxide a potent greenhouse gas. Ammonia also combines with other substances in the air to form fine particles, which can harm human health.

The UN/ECE Gothenburg Protocol, and the EU National Emissions Ceiling Directive have been implemented to control ammonia emissions (amongst other pollutants) at the national level. Both the Protocol and the Directive have national emission ceilings for 2010, and both are currently undergoing revision to include revised more stringent ceilings for 2020.

Several measures can be used to decrease the ammonia emissions from agricultural activities e.g. effective manure management from collection and storage to application to land. The most substantial reduction in emissions from manure management can be achieved by using different application techniques. The use of injection techniques (open slot or deep drilling) have been shown to provide substantially reduced emissions of ammonia. Trailing shoe and trailing hose techniques have also been shown to give reduced emissions, when compared to the dated splash-plate or broadcast application methods.

What advice is already available

In March 2009 AIC, FWAG, LEAF, NFU and CLA[2] released the Tried & Tested whole farm nutrient management plan, which states "Good nutrient management is one of the keys to farm profitability and reduced environmental risk." *Tried & Tested* is a paper–based plan which is intended to be used alongside the recommendations in this revised Fertiliser Manual.

---

2   Agriculture Industry Confederation (AIC); Farming and Wildlife Advisory Group (FWAG); Linking Environment and Farming (LEAF); National Farmers Union (NFU); Country Land and Business Association (CLA)

Introduction

Farm advisers who give crop nutrition advice are required to meet the latest standards of professional competence as determined by the FACTS Scheme. FACTS sets the professional standards in crop nutrition and is the body responsible for both setting and maintaining standards of advice given by individuals on farm. There are more than 2200 FACTS[3] Qualified Advisers (FQAs) across the UK.

Protecting our Water, Soil and Air: A Code of Good Agricultural Practice for farmers, growers and land managers, known widely as the 'CoGAP', consolidates and updates three separate former codes for water, soil and air. The publication offers practical interpretation of legislation and provides good advice on best practice: 'good agricultural practice' means a practice that minimises the risk of causing pollution while protecting natural resources and allowing economically sound agriculture to continue. The CoGAP helps farmers to think about the main operations that their type of farming business might undertake in terms of activities carried out in the field. It also helps farmers to consider the risks of what they do, and to develop and follow management plans for manures, plant nutrients, soils and crop protection.

Advice on improving water quality is currently available in 50 priority catchments in England through the England Catchment Sensitive Farming Delivery Initiative, (ECSFDI). This joint Defra/Natural England/Environment Agency project uses a network of Catchment Sensitive Farming Officers (CSFOs) to engage with farmers and encourage them to make changes in practice and infrastructure to improve water quality and help meet Water Framework Directive targets. The ECSFDI also operates a small but very popular Capital Grants Scheme to part fund infrastructure improvements on farms to help reduce diffuse water pollution.

The 'Nitrate Regulations'[4], which implement the European Community's Nitrates Directive, help reduce nitrogen losses from agriculture to water. They designate areas where nitrate pollution is a problem, known as Nitrate Vulnerable Zones (NVZs), and set rules for certain farming practices which must be followed within these zones.

The rules promote best practice in the use and storage of fertiliser and manure, and build on the guidelines set out in the CoGAP. The Government encourages farmers outside of the NVZs to follow this voluntary Code of Good Practice, in order better to protect the environment.

Defra and the Environment Agency have produced nine guidance leaflets to help farmers in NVZs understand the requirements, and implement and comply with the new Action Programme measures. These are available on the Defra website or via the NVZ Helpline.[5]

The PLANET 3 software generates recommendations based on this new Fertiliser Manual, and can generate records of nutrient use on the farm. It takes account of the crop nutrient requirement, the nutrients supplied from the soil and applications of organic manures and manufactured fertilisers. Defra make available the PLANET Dynamic Link Library, which performs the calculations to generate nutrient management recommendations, for incorporation into commercial software programmes to help with good nutrient management.

---

3 The Fertiliser Advisers Certification and Training Scheme (FACTS) was established in 1993 following discussions between the industry and BASIS (Registration) Ltd to provide a recognised standard of competence for UK advisers on crop nutrition and fertilizer use.
4 The Nitrate Pollution Prevention Regulations 2008' (SI 2008/2349).
5 www.defra.gov.uk/environment/quality/water/waterquality/diffuse/nitrate/help-for-farmers.htm

Section 1: Principles of nutrient management and fertiliser use

	Page
Crop nutrient requirements	16
Integrated plant nutrient management	17
Important soil properties	17
Soil acidity and liming	19
Nitrogen for field crops	22
Phosphorus, Potassium and Magnesium for field crops	33
Sulphur	43
Sodium	44
Micronutrients (trace elements)	45
Fertiliser types and quality	46
Fertiliser application	49
Protection of the environment	50

Section 1: Principles of nutrient management and fertiliser use

Crop nutrient requirements

Some 13 elements, in addition to carbon (C), hydrogen (H) and oxygen (O), are known to be essential for plant growth and they can be divided into two groups:

- Macronutrients: these are nitrogen (N), phosphorus (P), potassium (K), calcium (Ca), magnesium (Mg) and sulphur (S) and are required in relatively large amounts.
- Micronutrients (trace elements): these include iron (Fe), copper (Cu), manganese (Mn), zinc (Zn), boron (B), molybdenum (Mo) and chlorine (Cl), and are required in smaller amounts than the macronutrients.

The names macro- and micro- nutrients do not refer to relative importance in plant nutrition; a deficiency of any one of these elements can limit growth and result in decreased yield. It is therefore important to ensure that there is an optimum supply of all nutrients – if a plant is seriously deficient in, for example, potassium it will not be able to utilise fully any added nitrogen and reach its full potential yield and any unutilised nitrogen may be lost from the field.

In the UK, two conventions are used as follows:

- For fertiliser contents and for recommendations, phosphorus is expressed in the oxide form phosphate (P_2O_5) and potassium as potash (K_2O). Sulphur, magnesium and sodium also are expressed in oxide forms (SO_3, MgO and Na_2O).
- Soil and crop analysis reports usually show elemental forms for example mg P/kg or mg K/l.

Oxide or elemental forms are used in this *Manual* according to context.

Achieving the right timing of nutrient application is as important as applying the correct amount. Crop demand varies throughout the season and is greatest when a crop is growing quickly. Rapid development of leaves and roots during the early stages of plant growth is crucial to reach the optimum yield at harvest, and an adequate supply of all nutrients must be available during this time.

Excess application of nutrients, or application at the wrong time, can reduce crop quality and cause problems such as lodging of cereals or increases in foliar pathogens. Excessively large amounts of one nutrient in readily plant-available forms in the soil solution may also decrease the availability or uptake by the root of another nutrient.

Other elements found in plants, which may not be essential for their growth include, cobalt (Co), nickel (Ni), selenium (Se), silicon (Si) and sodium (Na). Sodium has a positive effect on the growth of a few crops. Some elements, such as cobalt, iodine (I), nickel and selenium are important in animal nutrition. These are normally supplied to the animal via plants, and must consequently be available in the soil for uptake by plant roots.

All these elements are taken up by plant roots from the supply in the soil solution (the water in the soil). They are absorbed in different forms, have different functions and mobility within the plant and hence also cause different deficiency, or very occasionally toxicity, effects and symptoms.

Section 1: Principles of nutrient management and fertiliser use

Integrated plant nutrient management

Crops obtain nutrients from several sources:

- Mineralisation of soil organic matter (all nutrients)
- Deposition from the atmosphere (mainly nitrogen and sulphur)
- Weathering of soil minerals (especially potash)
- Biological nitrogen fixation (legumes)
- Application of organic manures (all nutrients)
- Application of manufactured fertilisers (all nutrients)
- Other materials added to land e.g. soil conditioners

For **good nutrient management**, the total supply of a nutrient from all these sources must meet, but not exceed, crop requirement. Crop requirement varies with species (and sometimes variety of the crop), with yield potential (this in turn depends on soil properties, weather and water supply) and intended use (for example feed and milling wheat). Nutrients should be applied in organic manures or in fertilisers only if the supply from other sources fails to meet crop need. Where nutrients are applied, the amounts should be just sufficient to bring the total supply to meet crop need.

Matching nutrient supply to crop requirement involves several steps, some clear to the grower or adviser, others embedded in the recommendation system used. Recommendation methods used in this *Manual* allow the user to take account of sources of nutrients other than those applied, by using soil Index systems. In the case of phosphate and potash, recommendations are based on the expected crop yield. Using this *Manual* will help to match nutrient applications to crop need, maximising economic return and minimising costly nutrient loss to water and air.

Important soil properties

Soil texture

Knowledge of the soil type in the surface and subsoil of each field is essential for making accurate decisions on fertiliser and lime use. Without this knowledge it is not possible to use the recommendations in this *Manual* effectively and to achieve optimum benefit from them. Time is well spent therefore in acquiring and retaining this information because it is an intrinsic property of the soil that does not change with time. Soil type as used in this *Manual* is related to soil texture, which ranges from sands to clays. Soil texture is defined by the proportion of sand, silt and clay sized mineral particles in the soil and can be determined in a number of ways:

- Laboratory analysis of the proportion of the different mineral particles in the soil, followed by classification using the texture triangular diagram given in Appendix D of Controlling Soil Erosion (MAFF PB4093).

- Identification of the Soil Series for each field from the Regional Soil Maps for England and Wales, with classification from the accompanying Brown Book (available from the National Soil Resources Institute at Cranfield University).

Section 1: Principles of nutrient management and fertiliser use

- Assessment of texture class by hand using the method given in Appendix 1, followed by classification. Guidance also is given in think**soils** from the Environment Agency which helps farmers and farm advisers assess soil and recognise problems of erosion and runoff. Included is a pocket-sized quick guide to soil assessment to help the identification of soil texture.

Soil structure

To achieve optimum economic yields, crops have to acquire sufficient nutrients and water from the soil via the roots. It is therefore important to develop and maintain a good soil structure so that root growth is not adversely affected by poor physical soil conditions, such as compaction.

Soil mineral particles can be aggregated together and stabilised, either by clay or organic matter, to form crumbs. Within and between these crumbs are pores (voids) that can be occupied by air or water, both of which are required for roots to function properly. If the diameter of the pores is too small root tips cannot enter and roots cannot grow to find water and nutrients. If pores are too large water drains rapidly from the soil and roots will not grow because the soil contains too little moisture. The aggregation of soil mineral particles defines soil structure. For example, sands are often without any recognised structure while loamy soils can have an excellent structure. Developing and maintaining a good soil structure depends greatly on good soil management including cultivation at appropriate times and depths, and minimising traffic over the soil when it is too wet. *Protecting Our Soil, Water and Air: A Code of Good Agricultural Practice* contains general guidance on practices that will increase and maintain the ability of the soil to support plant growth.

Soil organic matter

Soil organic matter helps bind soil mineral particles of sand, silt and clay into crumbs. It has a number of important functions in crop nutrition. It improves soil structure enabling roots to grow more easily throughout the soil to find nutrients. It holds phosphorus and potassium ions (the forms taken up by roots) very weakly so that they are readily available for uptake by roots. It holds a store of organic forms of nitrogen, phosphate and sulphur from which available forms of these nutrients are released by microbial action.

The amount of organic matter in a soil depends on the farming system, the soil type and climate. The interplay between the first two factors is such that, in general, for the same farming system, a clay soil holds more organic matter than a sandy soil, and for the same soil type, a grassland soil holds more organic matter than an arable soil. It is difficult to define a critical level of soil organic matter because there are so many combinations of soil type and farming system. However, maintaining and where possible increasing soil organic matter should be a priority.

Soil mineral matter

The types of mineral matter in a soil affect the reserves of plant nutrients and their availability. For example, a soil containing minerals from igneous sources is likely to contain appreciable amounts of potash and possibly iron whereas a soil containing chalk will contain appreciable amounts of calcium. Soils overlying chalk are typically alkaline (pH above 7) while those overlying sandstone are typically acidic.

Section 1: Principles of nutrient management and fertiliser use

Stone content and rooting depth

A large content of impermeable stones increases the speed of water movement through the soil and because there is less fine soil to hold nutrients, the availability of water and nutrients is lower than in largely stone-free soils.

Soil rooting depth is important because many crops have root systems with a potential to grow to a metre deep or more. In deep friable soils where roots can grow to depth they can take up water and nutrients leached from the surface soil. Shallow soils over hard rock and compacted soil layers limit root growth, restricting nutrient and water availability and hence crop yields.

Soil acidity and liming

(See Section 6 for additional soil pH and liming recommendations for Fruit, vines and hops, and Section 8 for recommendations for grassland)

Soil pH is a measure of acidity or alkalinity. It can be measured in the laboratory using a soil sample taken from the field or directly in the field using a portable soil test kit. When determined in the laboratory, pH is usually measured in a soil/water suspension. The natural pH of a soil depends on the nature of the material from which it was developed. It ranges from about pH 4 (very acid), when most crops will fail, to about pH 8 for soils naturally rich in calcium carbonate (lime) or magnesium carbonate. For soils with a pH lower than 7, natural processes (e.g., rainfall, crop growth and especially leaching of calcium in drainage water) and some farming practices (e.g., use of large amounts of some nitrogen fertilisers) tend to acidify soil. Acidifying processes can cause soil pH to fall quite quickly and regular pH checks are advisable. Such acidifying processes rarely affect the pH of calcareous soils except perhaps in the top few centimetres where the soil is undisturbed. If problems are suspected, soil pH should be checked.

The optimum availability of most plant nutrients in soil occurs over a small range of soil pH values. Unfortunately the range for each nutrient is not the same but there is sufficient overlap in the ranges to decide the best possible compromise for each cropping system and soil type and these are shown below.

	Optimum soil pH[a]	
	Mineral soils	**Peaty soils**
Continuous arable cropping	6.5[b]	5.8
Grass with an occasional barley crop	6.2	5.5
Grass with an occasional wheat or oat crop	6.0	5.3
Continuous grass or grass/clover swards	6.0	5.3

a. The optimum pH is based on soil that has been correctly sampled (see Appendix 3). In some soil samples containing fragments of free lime, analysis of the ground soil sample in the laboratory can give a misleadingly high value for pH. The pH is measured in a soil/water suspension.

b. In arable rotations growing acid sensitive crops such as sugar beet, maintaining soil pH between 6.5 and 7.0 is justified.

Section 1: Principles of nutrient management and fertiliser use

Maintaining the optimum pH level in the topsoil in all parts of the field is important to achieve optimum yields and consistent quality. Not correcting soil acidity can cause large yield losses, but over-use of lime is wasteful and costly and can create problems with the availability of some micronutrients.

Lime recommendations

For each field the amount of lime to apply will depend on the current soil pH, soil texture, soil organic matter and the optimum pH needed. Clay and organic soils need more lime than sandy soils to increase pH by one unit. A lime recommendation is usually for a 20 cm depth of cultivated soil or a 15 cm depth of grassland soil. The table below gives examples of the recommended amounts of lime (t/ha of ground limestone or chalk, neutralising value NV 50-55) required to raise the pH of different soil types to achieve the optimum pH level shown in the table above.

Initial soil pH	Sands and loamy sands		Sandy loams and silt loams		Clay loams and clays		Organic soils		Peaty soils	
	Arable	Grass	Arable	Grass	Arable	Grass	Arable	Grass	Arable	Grass
					t/ha					
6.2	3	0	4	0	4	0	0	0	0	0
6.0	4	0	5	0	6	0	4	0	0	0
5.5	7	3	8	4	10	4	9	3	8	0
5.0	10	5	12	6	14	7	14	7	16	6

Where soil is acid below 20 cm, and soils are ploughed for arable crops, a proportionately larger quantity of lime should be applied. However, if more than 10 t/ha is needed, half should be deeply cultivated into the soil and ploughed down with the remainder applied to the surface and worked in. For established grassland or other situations where there is no, or only minimal, soil cultivation, no more than 7.5 t/ha should be applied in one application. In these situations, applications of lime change the soil pH below the surface very slowly. Consequently the underlying soil should not be allowed to become too acid because this will affect root growth and thus limit nutrient and water uptake, which will adversely affect yield.

Liming materials

The effectiveness of a liming material depends on its neutralising value (NV), its fineness of grinding and the hardness of the parent rock. The NV is the relative effectiveness of a liming material compared to that of pure calcium oxide (CaO). Lime recommendations are usually given in terms of ground limestone or ground chalk (NV 50-55), but other liming materials can be used provided allowance is made for differences in NV, fineness of grinding and cost.

The application rate is adjusted to take account of differences in NV and fineness of grinding of the materials because this affects the speed of reaction in the soil. *The Fertilisers Regulations*

Section 1: Principles of nutrient management and fertiliser use

(see Section 9) give details of the meaning and required declarations of different named liming products. In addition, materials such as sugar beet factory lime and lime treated sewage cake contain a useful amount of lime. Appendix 7 gives typical NVs of some common liming materials. The booklet *Agricultural Lime – the Natural Solution (Agricultural Lime Association)* gives more information on liming materials.

The cost of different liming materials can be compared by calculating the cost per unit of NV but allowance should also be made for any differences in particle fineness.

> **Example**
>
> *Ground limestone has an NV of 50 and costs £20/t delivered and spread. An alternative liming material (A) has an NV of 30 and costs £17/t delivered and spread.*
>
> Ground limestone costs (20 x 100) / 50 = 40 pence per unit of NV.
> Liming material A costs (17 x 100) / 30 = 57 pence per unit of NV.
>
> Provided the two materials have the same physical characteristics, the ground limestone is the more cost-effective liming material.

Some liming materials contain other useful nutrients and this also should be taken into account when deciding on which to use. For example, magnesian limestone (dolomitic limestone) contains large amounts of magnesium and is effective for correcting soil magnesium deficiency as well as acidity. However, many years of using magnesian limestone can result in an excessively high soil Mg Index and excess magnesium in the soil may induce potash deficiency in crops.

Lime application

It is important to maintain the appropriate soil pH for the cropping system and soil type and soil pH should not vary by more than ± 0.5 pH unit from the optimum. However, when to apply lime can be fitted in with the crops being grown. For example:

- Potatoes can tolerate acid soils and applying lime immediately before potatoes are grown should be avoided because of the risk of increased levels of potato scab.

- Sugar beet and barley are sensitive to soil acidity. Lime should be applied before these crops are grown.

- Clover is more sensitive to soil acidity than are many grass species and soil pH should be maintained to encourage a clover-rich sward.

A liming material should always be well worked into the cultivated soil because it can take some months to have its full beneficial effect in increasing pH throughout the topsoil. It is unwise to grow a crop which is sensitive to acidity immediately after liming a very acid soil. If it is important to try to achieve a rapid effect then the use of a fast acting liming material could be considered.

Section 1: Principles of nutrient management and fertiliser use

Nitrogen for field crops

(Section 8 contains additional information for grassland)

Most agricultural soils contain too little, naturally occurring plant-available nitrogen to meet the needs of a crop throughout the growing season. Consequently, supplementary nitrogen applications have to be made each year. Applying the correct amount of nitrogen at the correct time is an essential feature of good crop management.

Crop nitrogen requirement

'Crop nitrogen requirement' is the amount of nitrogen that should be applied to give the on-farm economic optimum yield. Nitrogen recommendations for all crops in this *Manual*, except grass, are crop nitrogen requirements defined in this way. Crop nitrogen requirement should not be confused with total nitrogen uptake by the crop or with the total supply of nitrogen (including that from the soil) that is needed by the crop. For grassland the *new systems based approach* described above provides recommendations based on the economic need to produce the amount of home grown forage necessary to maintain a target intensity of production, rather than the on-farm economic optimum of just the grass crop.

Basis of the Recommendations

Provided there are adequate supplies of water and other nutrients, nitrogen usually has a large effect on crop growth, yield and quality. The diagram below shows a typical nitrogen response curve. Applying nitrogen gives a large increase in yield but applying too much can reduce yield by aggravating problems such as lodging of cereals, foliar diseases and poor silage fermentation. When too much nitrogen is applied, a larger proportion is unused by the crop. This is a financial cost and can also increase the risk of nitrate leaching to water and contribute to other environmental problems such as climate change: Appendix 11 explains in more detail these other costs of fertiliser use.

Section 1: Principles of nutrient management and fertiliser use

A Typical Nitrogen Response Curve

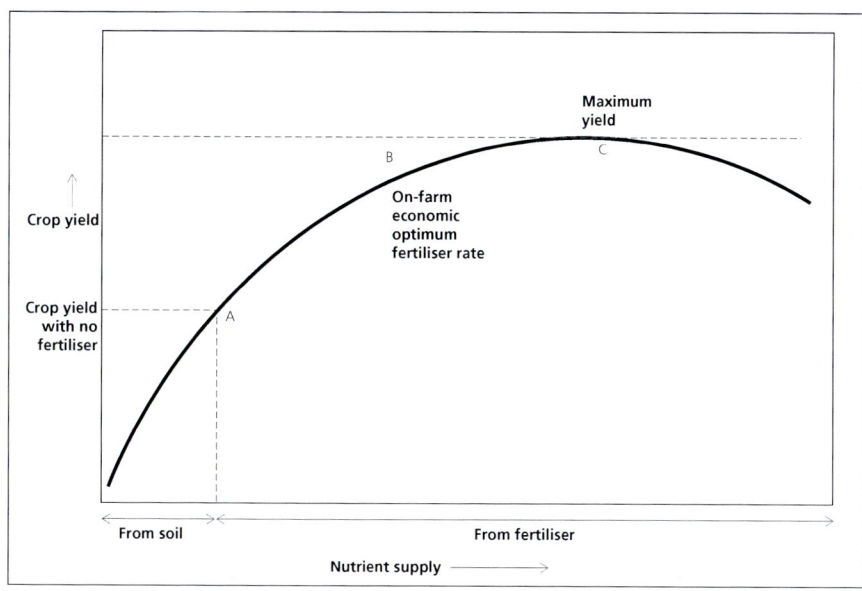

With reference to diagram above:

- Without applied nitrogen, yield typically is low (A).
- As nitrogen use increases from very small amounts, there is a large increase in yield up to the '*on-farm* economic optimum' nitrogen rate (B). This rate depends on the cost of the applied nitrogen and on the value of the crop ('breakeven ratio') as well as on the shape of the response curve. **All recommendations in this Manual are calculated using a typical breakeven ratio (see page 32) to provide the best on-farm economic rate of nitrogen to apply**. Substantial changes in the value of the crop produce or in the cost of nitrogen are needed to alter the recommendations. Where appropriate, different recommendations are given to achieve crop quality specifications required for different markets. Appendix 11 describes on-farm economic optimum (or profit maximisation) in more detail.
- Application of nitrogen above the on-farm economic optimum will increase yield slightly but this yield increase will be worth less than the cost of the extra nitrogen.
- Maximum yield (C) is reached at a nitrogen rate greater than the on-farm economic optimum and this is never a target if farm profits are to be maximised. Application of nitrogen above point C does not increase yield, and with further applications yield falls and the need for agro-chemicals such as fungicides and growth regulators may increase.
- **The nitrogen can be supplied from fertiliser and/or organic manure.**
- At nitrogen rates up to the on-farm economic optimum, there is a roughly constant amount of nitrogen left in the soil at harvest. At nitrogen rates above the on-farm economic optimum, there will be a larger surplus of residual nitrogen, usually as nitrate, in soil after harvest. This nitrate is at risk of loss in ways that can cause environmental problems like leaching to ground or surface water and denitrification to nitrous oxide (a greenhouse gas). For this reason, the amount of nitrate leached begins to increase by larger amounts above the on-farm economic optimum. There are social costs associated with these environmental problems, and this is explained in more detail in Appendix 11.

Section 1: Principles of nutrient management and fertiliser use

Nitrogen supply and losses

It is convenient to think about nitrogen in terms of 'nitrogen supply' and 'nitrogen losses'. Nitrogen supply can be from the soil, the atmosphere and organic manures as well as from fertiliser. Nitrogen losses may be by leaching, run-off, ammonia volatilisation and denitrification.

Nitrogen supply

Soil Mineral Nitrogen (SMN)	Nitrate-N (NO_3-N) and ammonium-N (NH_4-N) are often called mineral nitrogen. Both are potentially available for crop uptake and the amount in the soil depends on the recent history of cropping, organic manure and nitrogen fertiliser use.
Nitrogen mineralised from organic matter	Mineralisation results in the conversion of organic nitrogen to mineral nitrogen by soil microbes. The amount of organic nitrogen mineralised can be large: • on organic and peaty soils • where organic manures have been used for many years • where nitrogen-rich, organic material is ploughed into the soil.
Nitrogen from the atmosphere	Small amounts of nitrogen are deposited in rainfall and directly from the atmosphere. Leguminous crops, like peas, beans and clover, have bacteria in the nodules on the roots that can 'fix' atmospheric nitrogen into a form that can be used by the plant.
Nitrogen from organic manures	Most organic manures contain some mineral nitrogen, which is equivalent to mineral nitrogen in fertilisers. The remaining organic nitrogen becomes available more slowly (see Section 2).
Manufactured fertiliser nitrogen	Fertiliser nitrogen is used to make up any shortfall in the crop's requirement for nitrogen.

Nitrogen losses

Leaching	Nitrate is soluble in the soil solution and, unlike ammonium, is not held on soil particles. Once the soil is fully wetted, nitrate may leach into field drains or sub-surface aquifers as drainage water moves through the soil. Leaching is more rapid on light sand soils compared to deep clay or silt soils which are less free draining and therefore more retentive of nitrate. The amount of winter rainfall has an important influence on the amount of nitrate leached. Although ammonium-N can be strongly fixed to clay particles and is less at risk to leaching than nitrate, under normal conditions ammonium-N in the soil is rapidly converted to nitrate. In practice, sources of nitrogen containing or converted to ammonium-N will have a similar risk of leaching as sources containing nitrate when used in excess of the requirement of a crop.

Section 1: Principles of nutrient management and fertiliser use

Run-off	During and following heavy rainfall, nitrogen in solution or in organic form can move across the soil surface and enter watercourses. The amount of nitrogen lost from soil in this way will vary widely from field to field and season to season depending on the amount and timing and intensity of rainfall and nitrogen applications. Sloping ground, proximity to surface waters and application of slurry present particular risks of nitrogen loss in run-off.
Denitrification	In anaerobic soils (poorly aerated soils lacking oxygen), nitrate can be denitrified and lost to the atmosphere as the gases nitrous oxide, a green-house gas, and nitrogen (N_2). Denitrification is a biological process and is most significant in wet and warm soils where there is a supply of nitrate after harvest or where there has been a recent nitrogen application and there is enough organic matter for the microbes to feed on. Some nitrous oxide is formed during nitrification of ammonium-N to nitrate-N and some of this also can be lost to the atmosphere.
Ammonia volatilisation	Nitrogen may be lost to the atmosphere as ammonia gas. Significant losses commonly occur from livestock housing, livestock grazing and where organic manures are applied to fields and are not immediately incorporated by cultivation. There can also be significantly larger losses of ammonia when urea is applied compared to losses when other forms of nitrogen fertiliser, such as ammonium nitrate are used.

Factors influencing decisions about nitrogen use

The crop nitrogen requirement (the 'on-farm economic optimum') will depend on:

- The amount of nitrogen from all sources, including the soil, which must be available to achieve the optimum on-farm economic yield.
- The amount of nitrogen that the soil can supply for crop uptake.
- The cost of nitrogen fertiliser and the likely value of the crop.
- Any particular crop quality requirements, for example grain protein in bread making wheat or in malting barley.

In addition to identifying crop nitrogen requirement as accurately as possible, it may be necessary to comply with regulatory restrictions on the amount or timing of applications, for instance in NVZs.

When calculating how much manufactured fertiliser nitrogen to use, all supplies and losses of nitrogen, and the efficiency of fertiliser nitrogen use by the crop must be considered.

Section 1: Principles of nutrient management and fertiliser use

Assessing the Soil Nitrogen Supply (SNS)

For the purposes of using this *Manual*, the Soil Nitrogen Supply is defined as follows:

"The Soil Nitrogen Supply (SNS) is the amount of nitrogen (kg N/ha) in the soil (apart from that applied for the crop in manufactured fertilisers and manures) that is available for uptake by the crop throughout its entire life, taking account of nitrogen losses."

The SNS is different to, but includes Soil Mineral Nitrogen (SMN). The calculation of SNS must include three separate components of nitrogen supply as follows.

> **Soil Nitrogen Supply (SNS) = SMN + estimate of nitrogen already in the crop + estimate of mineralisable soil nitrogen**
>
> where:
>
> - Soil Mineral Nitrogen (kg N/ha) is the nitrate-N plus ammonium-N content of the soil within the potential rooting depth of the crop, allowing for nitrogen losses.
> - Nitrogen already in the crop (kg N/ha) is the total content of nitrogen in the crop when the soil is sampled for SMN.
> - Mineralisable soil nitrogen (kg N/ha) is the estimated amount of nitrogen which becomes available for crop uptake from mineralisation of soil organic matter and crop debris during the growing season after sampling for SMN.

The SNS depends on a range of factors which commonly vary from field to field and from season to season. It is therefore important to assess the SNS for each field each year. The key factors influencing SNS are:

- Nitrogen residues left in the soil from fertiliser applied for the previous crop.
- Nitrogen residues from any organic manure applied for the previous crop and in previous seasons.
- Soil type and soil organic matter content.
- Losses of nitrogen by leaching and other processes (the amount of winter rainfall is important).
- Nitrogen made available for crop uptake from mineralisation of soil organic matter and crop debris during the growing season.

Mineral nitrogen residues after harvest

The management and performance of a crop can have a significant effect on the amount of residual mineral nitrogen (nitrate-N and ammonium-N) in the soil at harvest. Residues are likely to be small if the amount of nitrogen applied matched crop demand in high yielding years or where the amount of nitrogen applied was less than that required by the crop. The residues may be larger than average when yields are unusually small due to disease or drought. Residues following cereals are generally lower than those following break crops.

Section 1: Principles of nutrient management and fertiliser use

Well-established cover crops, such as mustard, forage rape or *Phacelia*, sown after harvest can take up significant amounts of soil mineral nitrogen and reduce the risk of nitrate leaching over winter. Generally, the earlier the cover crop can be established, the more mineral nitrogen will be taken up. Following destruction of the cover crop, this nitrogen will be gradually mineralised over many years. However, the amount becoming available for uptake by the next crop is relatively small and difficult to predict. Where cover crops have been used regularly, soil analysis can be a useful technique to help estimate the overall supply of soil mineral nitrogen.

In ley-arable rotations, the nitrogen released from grass leys may persist for up to three years following ploughing, but most useful nitrogen becomes available within the first one or two seasons.

Effect of excess winter rainfall

The amount of nitrate leached will depend on the quantity in the soil when the water content reaches field capacity and through-drainage starts, the soil type and the amount of water draining through the soil (the excess winter rainfall).

The excess winter rainfall is the actual rainfall between the time when the soil profile becomes fully wetted in the autumn (field capacity) and the end of drainage in the spring, less evapo-transpiration during this period (i.e. water lost through the growing crop). In England and Wales, typical total evapo-transpiration between October and February inclusive is around 50mm with a further 28mm in March. The Met Office can provide estimates of excess winter rainfall in different locations and for different soil/cropping situations.

> **Excess winter rainfall (mm) = Rainfall between the time a soil reaches field capacity and the end of drainage – evapo-transpiration**

Light sand soils and some shallow soils can be described as 'leaky'. Nitrate in these soils following harvest is fully leached in an average winter even where substantial residues are present in the autumn. The SNS Index is nearly always 0 or 1 and is independent of previous cropping except in low rainfall areas or after dry winters.

Deep clay and silt soils can be described as 'retentive'. The leaching process is much slower and more of the nitrate residues in autumn will be available for crop uptake in the following spring. Differences in excess winter rainfall will have a large effect on SNS in these soils. Low levels of SNS (Index 0 and 1) are less frequent than on sandy soils. Other mineral soil types are intermediate between these two extremes.

Because of both regional and seasonal differences, separate SNS Index tables are given for three different rainfall situations (see Section 3).

1. Up to 600 mm annual rainfall (up to 150 mm excess winter rainfall)
2. 600-700 mm annual rainfall (150-250 mm excess winter rainfall)
3. Over 700 mm annual rainfall (over 250 mm excess winter rainfall)

Section 1: Principles of nutrient management and fertiliser use

Nitrogen released from mineralisation of organic matter

Nitrogen is released in mineral forms when microbial action breaks down soil organic matter. The rate of mineralisation depends on temperature and usually is small over winter until soil temperature reaches around 4°C. Small amounts of nitrogen may need to be applied in early spring for crops that have a significant nitrogen requirement at this time. In organic and peaty soils, mineralisation of soil organic matter in late spring and summer results in large quantities of nitrate becoming available for crop uptake.

Soil temperature often remains higher than 4°C for some weeks after harvest supporting mineralisation in autumn that can contribute to the nitrogen requirement of an autumn-sown crop or to the risk and extent of nitrate leaching where the land remains uncropped.

Long season crops (e.g. sugar beet) will utilise more mineralised nitrogen than crops which are harvested in mid- or late-summer. For example, cereals make little use of nitrogen mineralised after June.

For the purposes of this *Manual*, organic soils are considered to contain between 10 and 20% organic matter. In such soils, the amount of SNS is likely to vary widely depending on the amount and age of this soil organic matter. However, the relationship between the actual SNS and the soil organic matter level is poor. Some soils with more than 10% organic matter can have a SNS similar to that of mineral soils. The recommendations in the tables for organic soils are based on an organic matter content of 15%.

Peaty soils contain over 20% organic matter. They are always at SNS Index 5 or 6 irrespective of previous cropping or manuring history, or excess winter rainfall. This is because the large amounts of organic nitrogen mineralised will usually be much greater than variations in the nitrogen residues due to previous cropping.

The amount of nitrogen mineralised from past applications of organic manures (over 1 year old) is difficult to estimate. The amount will generally be small. It can be greater where there has been a history of large regular applications of organic manures and in these situations it can be worthwhile to sample the soil and analyse it for SMN.

Organic nitrogen in crop debris from autumn harvested crops usually mineralises quickly and nitrate is liable to loss by leaching over winter in the same way as mineral nitrate residues from fertilisers. Mineralisation of nitrogen-rich leafy debris is quicker than that of nitrogen-poor straw debris. Organic nitrogen that is not mineralised quickly becomes available over a long time and may contribute little to the nitrogen supply of the following crop. Examples of cropping situations where mineralised nitrogen from crop residues can make a useful contribution are:

- Incorporation of sugar beet tops, especially before a late autumn-sown crop.
- Where a second cauliflower crop is grown in the same season. Large amounts of leafy, nitrogen-rich crop debris will be returned to the soil after harvest of the first crop, and will quickly release nitrate available for the next crop.

Section 1: Principles of nutrient management and fertiliser use

The Soil Nitrogen Supply (SNS) Index System

The nitrogen recommendations in this *Manual* are based on seven SNS Indices and each Index is related to a quantity of SNS in kg N/ha. The SNS Index can be determined using either field specific information (the 'Field Assessment Method') without sampling and analysis for SMN, or by using the results of soil sampling and analysis for SMN and an assessment of any nitrogen already taken up by the crop (the 'Measurement Method').

A nitrogen recommendation is obtained by determining the SNS Index of the field using one of these methods, then referring to the appropriate crop table to obtain the nitrogen recommendation for the selected Index. The SNS Index system is not applicable for established grassland or established fruit crops.

Note: The Field Assessment Method for SNS Index does NOT take account of organic manure applied since the last crop was harvested, or any manure that will be applied during the growing season. The Measurement Method does not take account of any manures applied after the soil sample for SMN was taken. The nitrogen contribution from such manure application must be calculated separately (see Section 2).

Full details of the SNS Index system and how to use it (with examples) are given in Section 3:

The Field Assessment Method

In most situations the SNS Index will be identified using the Field Assessment Method which is based on field specific information for previous cropping, previous fertiliser and manure use, soil type and winter rainfall. The SNS Index is read from tables and there is no requirement for soil sampling or analysis.

For the Field Assessment Method, it is assumed that all previous crops have been managed well and that previous nitrogen fertiliser use has been close to the recommended rate, taking account of the application of any organic manure.

This method can provide a satisfactory assessment of the SNS in typical arable rotations but the Measurement Method may give a better result where SNS could be outside the normal range.

The Measurement Method

SNS is related to the cumulative effects of farming practice over several years and can be difficult to predict in some circumstances. Thus, soil sampling and analysis for SMN with an assessment of nitrogen already taken up by the crop and of the contribution of mineralisable nitrogen is recommended where the soil to rooting depth may contain large or uncertain amounts of plant-available nitrogen. This method should be targeted to fields where the supply of plant-available nitrogen in the soil could be unusually large, and particularly where organic manures have been used regularly in recent years.

In these situations, direct measurement and estimation of the key components of SNS (SMN, total crop nitrogen content and mineralisable nitrogen) will usually result in the best assessment of the amount of soil nitrogen available for a crop and lead to a more accurate decision about how much additional nitrogen to apply.

Section 1: Principles of nutrient management and fertiliser use

Soil Nitrogen Supply (SNS) = SMN + estimate of total crop nitrogen content + estimate of mineralisable nitrogen

Soil sampling and analysis for SMN is the most important component and should be carried out carefully using the procedures recommended in Appendix 2. Sampling and analysis for SMN is NOT recommended on peaty soils or in the first year after ploughing out grassland. This is because the mineralisation of organic nitrogen after the soil has been sampled can produce a large amount of plant-available nitrogen and this becomes a major part of the SNS, which is not measured. Also the measurement of SMN is not recommended for use in established grassland systems.

Note: The nitrogen contribution from manures applied before sampling for SMN will be largely taken account of in the measured value and should not be calculated separately.

To identify the SNS Index of the soil when the Measurement Method is used, it is important to add to the SMN result an estimate of crop nitrogen content and, where necessary, the mineralisable nitrogen.

The total crop nitrogen content of cereals and oilseed rape at the time of SMN sampling can be estimated using the scheme given in Appendix 2. No similar scheme is available for estimating the nitrogen content of other crops at this time of year.

It is much more difficult to obtain a reliable estimate of the nitrogen that will be made available from mineralisation of organic matter. However on many mineral soils, this will be small and can be ignored for all practical purposes. On organic and peaty soils, or where large amount of organic material (crop residues or organic manure) have been applied in recent years large quantities of nitrogen can be mineralised.

Where shallow rooted crops are to be grown (e.g. some vegetables) and soil sampling for SMN is not done to the full 90 cm depth, the results of analysis must be adjusted to determine the SNS Index. Uniform concentration of mineral nitrogen down the soil profile to 90cm can be assumed though this will introduce some error. Some existing decision support systems can be used to interpret these results to provide a fertiliser recommendation (e.g. WELL_N).

Nitrogen Uptake Efficiency by Crops

The efficiency of uptake of nitrogen from different sources varies, even for well-grown crops.

Provided that the total soil nitrogen supply does not exceed demand, most crops will take up an amount of soil nitrogen that is roughly equivalent to the amount of SMN present within rooting depth. Thus, for every 50 kg N/ha of SMN that is measured, the crop is likely to take up approximately 50 kg N/ha, i.e., SMN is used with apparently 100% efficiency.

The actual efficiency with which SMN is recovered is likely to be less than 100% (and might typically be closer to 60%). However, this will generally be compensated for by additional soil nitrogen that becomes available for uptake during the growing season (often through mineralisation of crop residues and soil organic matter).

Section 1: Principles of nutrient management and fertiliser use

The efficiency with which SMN is recovered is likely to be lower where a large amount is present, especially in the autumn at below topsoil depth and/or on sandy soils or in high rainfall situations where nitrate leaching losses will be greater. Early establishment of crops may help to reduce losses and increase the uptake of SMN.

The application of amounts of nitrogen in line with crop demand does not appear to decrease the uptake of soil nitrogen. However, for crops suffering from the adverse effects of disease, poor soil conditions, drought or other growth inhibiting problems, the efficiency of uptake of both fertiliser and soil-derived nitrogen is often reduced.

Research has shown that the efficiency of uptake of fertiliser nitrogen by winter wheat and winter barley varies depending on the soil type. The values shown below are largely based on work with ammonium nitrate, and values for other types of nitrogen fertiliser may be different in some circumstances.

Light sand soils	70% efficiency*
Medium, clay, silty, organic and peaty soils	60% efficiency
Shallow soils over chalk and limestone	55% efficiency

* i.e., 70 kg N/ha taken up by the crop for every 100 kg N/ha applied as fertiliser.

The winter wheat and winter barley recommendations in this book are adjusted to take account of these differences in nitrogen fertiliser uptake efficiency due to soil type. There has been insufficient research to show if these differences can be applied to other crops.

Timing of Nitrogen Applications

Correct timing of nitrogen fertiliser application is important so that crops make best use of the nitrogen applied and there is minimum risk of adverse environmental impact of the application. As a general principle, nitrogen should be applied at the start of periods of rapid crop growth and nitrogen uptake. The diagram below shows the typical pattern of nitrogen uptake by a winter wheat crop. It is easy to see why there is no benefit from applying autumn nitrogen to winter cereal crops. The nitrogen requirement is small during the autumn and winter and the supply from soil reserves is adequate to meet the requirement. However, autumn nitrogen is recommended for some winter oilseed rape crops reflecting the larger requirement for nitrogen of this crop in autumn.

The chart for the nitrogen uptake by a winter cereal shows that:

- In autumn/winter (A), there is only a small crop nitrogen requirement that can easily be met by soil nitrogen reserves. There is no need to apply nitrogen in autumn.

- The main period of nitrogen uptake (B) is March-June and during this growth period, there is usually insufficient soil nitrogen to support unrestricted growth. Nitrogen fertiliser should be applied at the start and during this period of growth.

Section 1: Principles of nutrient management and fertiliser use

Nitrogen Uptake by a Winter Cereal Crop in Relation to Available Soil Nitrogen

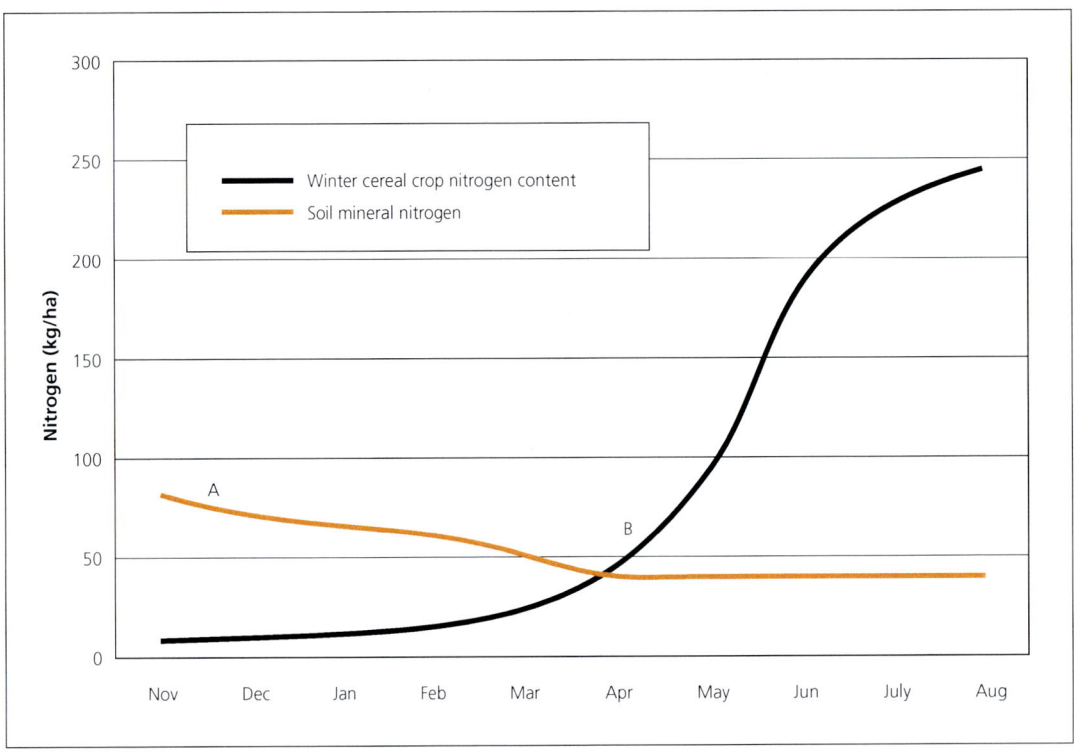

Nitrogen timing can also have a range of other important effects on crop growth and quality.

- Too much seedbed nitrogen can reduce the establishment of small seeded crops.
- Early spring nitrogen will increase tillering of cereals. This may be beneficial, but too much nitrogen at this stage can increase the risk of lodging.
- Late applied nitrogen will increase the grain nitrogen/protein concentration of cereals.

Recommended nitrogen timings for individual crops are given in the recommendation tables. However NVZ rules relating to closed periods take priority and may also influence timing of nitrogen applications.

The effect of economic changes

As a general principle, the recommendations are insensitive to changes in the value of the crop produce or the cost of nitrogen fertiliser. It would normally require a large change in the ratio of the value of a crop and the cost of nitrogen fertiliser to alter the recommendation.

The breakeven ratio is the crop yield (kg) needed to pay for 1 kg of nitrogen. If there are large changes in the ratio (the breakeven ratio) it will be appropriate to make adjustments in the recommendations.

$$\text{Breakeven ratio} = \frac{\text{Cost of nitrogen (pence/kg)}}{\text{Value of crop produce (pence/kg)}}$$

Section 1: Principles of nutrient management and fertiliser use

> **Example**
>
> Ammonium nitrate (34.5% N) costs £250/t or $\dfrac{250 \times 100}{34.5 \times 10}$ = 72 pence/kg N
>
> Wheat is sold for £120/t or $\dfrac{120 \times 100}{1000}$ = 12.0 pence/kg
>
> The breakeven ratio is $72 \div 12.0 = 6$
>
> The prices can be expressed in £/kg as long as both are in the same units (p/kg or £/kg).

The nitrogen recommendation tables for wheat, barley and oilseed rape show how to calculate adjustments to take account of the cost of nitrogen and the selling price of the grain produced.

Alternative approaches to nitrogen decisions

There is ongoing research into new approaches and techniques for deciding on nitrogen fertiliser use for cereals. The canopy management technique and use of spectral reflectance sensors are two examples. Such methods are increasingly being used to assess nitrogen requirement in the field. Other fertiliser nitrogen recommendation systems are based on models of crop growth and nitrogen uptake or on soil sampling to shallower depths than 90cm. Some include a measurement of mineralisable nitrogen in a soil sample.

Provided a recommendation system takes proper account of the total amount of nitrogen needed by a crop and of the supplies available from the soil and organic manures, it should give a recommendation close to crop nitrogen requirement.

Phosphate, Potash and Magnesium for field crops

(Section 8 contains additional information for grassland)

Phosphate, potash and magnesium applied in fertilisers and manures move only slowly through the soil and many soils can hold large quantities of these nutrients in forms that are readily available for crop uptake over several years. Consequently managing the supply of these three nutrients for optimum yield is based more on maintaining appropriate amounts in the soil for the needs of the rotation than on those of an individual crop. In practice, this means maintaining target soil Indices that ensure optimal phosphate, potash and magnesium nutrition.

As the amount of crop-available phosphate or potash in the soil increases from a very low level, crop yield increases, rapidly at first then more slowly until it reaches a maximum. Typically, maximum yield of arable crops or of grass is reached at Index 2 for phosphorus and Index 2- for potassium (see *Target Soil Indices* below). The principle for phosphate and potash management is to maintain the soil at the appropriate target Index. If the Index is lower than the target, yield may be reduced and additional phosphate and potash should be applied. If the Index is higher than the target, applications can be reduced or omitted until the soil level

June 2010

Section 1: Principles of nutrient management and fertiliser use

falls to the target Index. Effective use of target Indices depends on representative soil sampling. If it is felt that significant areas of the field could differ in P or K Index, these areas should be sampled and treated separately.

To maintain soils at the correct Index it is usually sufficient to replace the amount of each nutrient removed from the field in the harvested crop. This amount can be calculated from the yield and an average concentration of the nutrient in the harvested product as shown in Appendix 5 (see worked example on page 38). To check that this approach is maintaining the required P or K Index (i.e. the phosphate and potash status of the soil), soil sampling should be carried out every 3-5 years, at a suitable time in the crop rotation. Maintaining the appropriate level of phosphate and potash in the soil is especially important as these nutrients move slowly in the soil to plant roots. Once deficiency has occurred, a fresh application of phosphate and potash is most unlikely to be available for uptake by roots in time to benefit the crop being grown.

Yield response of some crops to soil phosphorus and potash concentrations

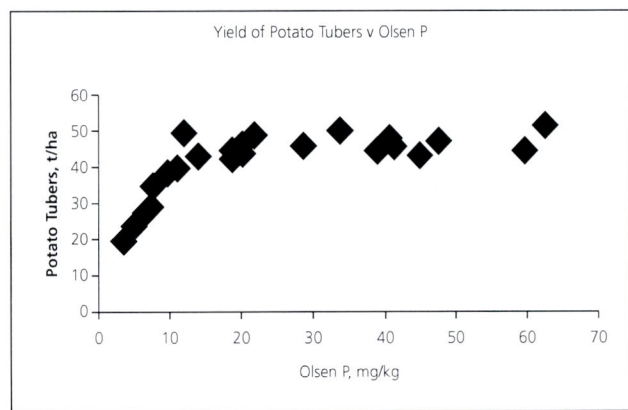
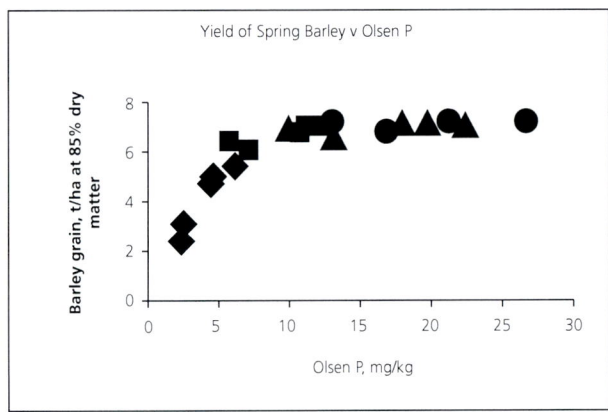
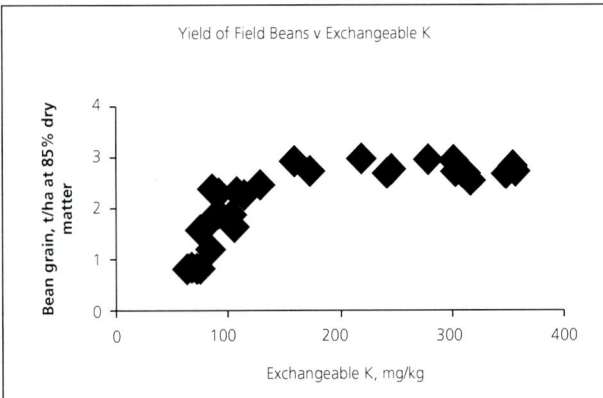
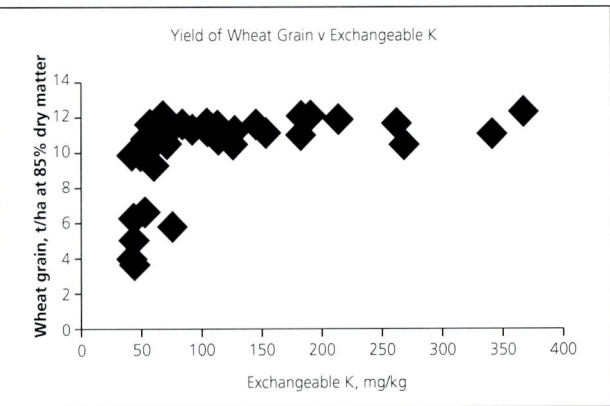

Section 1: Principles of nutrient management and fertiliser use

Developing an approach to manage soil phosphate, potash and magnesium supply

An appropriate approach for managing phosphate, potash and magnesium to maintain soil fertility must take account of:

- The current readily crop-available status of the soil in the field.
- The target Index or critical level of each nutrient for the rotation.
- The need to build up or maintain the target Index or possibly run it down.
- The responsiveness of a crop to a fresh application of each nutrient.
- The quantity of each nutrient removed from the field in crop produce.
- The quantity of nutrients supplied from any organic manure that is available for application (see Section 2).

Recommendations for phosphate and potash applications to build up soils in Index 0 and 1 and maintenance applications for soils at Index 2 are given in Sections 4 to 8 for different crops. Guidance on how to use the recommendation tables is given in Section 3: **The recommendations in this book are given as kg/ha of P_2O_5, K_2O and MgO. This is because the concentrations of phosphorus and potassium are expressed in this way in fertilisers and this makes it easy to calculate the amount of fertiliser to apply.** For example, if the recommendation is to apply 50 kg P_2O_5/ha and the fertiliser contains 20% P_2O_5, 50÷20 x 100 = 250 kg/ha of the fertiliser. Conversion tables (metric-imperial; element-oxide) are given in Appendix 8.

Soil sampling and analysis

Good management of soil phosphate, potash and magnesium depends on regular soil sampling and analysis. Levels of these nutrients in the soil change only slowly so soil sampling and analysis can be done every 3-5 years at an appropriate time in the crop rotation. It is usually safe to use soil analysis results for phosphorus, potassium and magnesium as a basis for fertiliser recommendations for up to 4 years from the date of sampling.

The analytical results will be meaningful only if an adequate and representative soil sample is taken. The recommended procedure for soil sampling is described in Appendix 3. For advisory purposes the results are usually given as milligrams of phosphorus, potassium and magnesium per litre of soil (mg P/litre, mg K/litre, and mg Mg/litre). The results are also given as an Index and the Indices range from 0 to 9 (see Appendix 4 for details). The Index system is based on the likely response of a crop to a fresh application of the nutrient because there is a good relationship between these two factors. Thus Index 0 soils are deficient and there would be an increase in yield by applying phosphate or potash. As the Index number increases the response to a fresh application of fertiliser declines and for most soils growing arable crops there would be no or only a very small response at Index 2. In Appendix 4, the Index number increases to 9 to accommodate glasshouse soils.

The fertiliser recommendations given in this book are based on the results of soil analysis using the following standard laboratory methods. These methods, which are well-tried and tested over many years, are appropriate for soils in England and Wales.

June 2010

Section 1: Principles of nutrient management and fertiliser use

Phosphorus (P)	measured in a sodium bicarbonate soil extract at pH 8.5 (Olsen's method, Olsen P)
	or
	measured by equilibration with an anionic resin in a soil suspension (resin P)
Potassium (K) and Magnesium (Mg)	measured in an ammonium nitrate soil extract (exchangeable K and exchangeable Mg)

Full details of the analytical methods are given in *Specification for topsoil and requirements for use. British Standard BS 3882: 2007* or *The Analysis of Agricultural Materials (MAFF RB427)*. The resin P analysis method is described by *Hislop J and Cooke I J (1968) Anion exchange resin as a means of assessing soil phosphate status: a laboratory technique. Soil Science,* **105**, *8-11.*

Potash-releasing clay soils

Depending on the nature of the minerals in the parent material from which they were developed, some heavy clay soils contain large quantities of potash. Weathering (breakdown) of these minerals releases the potash, which gradually becomes available for crop uptake. Unfortunately no routine soil analysis method is available to estimate the amount or rate of release of this potash. However, the potash that is released goes to the readily crop-available pool measured by soil analysis so that this value does not decline quickly if the amount of potash applied is less than that removed in the harvested crops. Local knowledge and past experience can be useful when assessing the potash release characteristics of clay soils. If the crop-available potash status of a clay soil changes little when the potash balance is consistently negative over a number of years this is a useful indicator that potash is being released from the clay by weathering. Remember that the annual rate of potash release may not be sufficient to meet the requirement of crops with a large yield potential requiring large amounts of potash. It is essential to monitor crop yields to ensure that the yield potential of the site is being realised and if this is not the case then potash fertilisers should be applied. A rough classification of clays based on their likely potash release characteristics is as follows:

Potash-releasing clays.	Chalky boulder clay, Gault clay, Weald clay, Kimmeridge clay, Oxford clay, Lias clay, Oolitic clay
Clays which do not release much potash.	Carboniferous clay

Target Soil Indices

The readily crop-available pool of these nutrients is measured by the analytical methods given above. The critical value for all three nutrients is related to a target Index. Fertiliser and manure applications should aim to raise the Index to the appropriate target for the rotation and then to maintain this target by maintenance applications.

Section 1: Principles of nutrient management and fertiliser use

The target soil Indices for soil P and K are given below.

	Soil P	Soil K
Arable and forage crops, grassland	Index 2 (16-25 mg/litre)	Index 2- (121-180 mg/litre)
Vegetables	Index 3 (26-45 mg/litre)	Index 2+ (181-240 mg/litre)

Where crops are grown on soils below the target Index applying large amounts of phosphate and potash rarely produces yields equal to those where the crop is grown on soil at the target Index. This is particularly likely where soil P or K Index is 0 as shown in the following examples taken from data in the Rothamsted Research archive.

Yields of spring barley and sugar beet on soils at different P Indices with and without freshly applied phosphate

		P Index			
		0	1	2	3
		mg P/L			
		5	12	20	30
Grain yield t/ha	No added P	3.15	4.60	5.10	5.40
	125 kg P_2O_5/ha added	4.20	5.00	5.25	5.40
Sugar yield t/ha	No added P	2.80	5.45	6.25	6.65
	125 kg P_2O_5/ha added	5.40	6.10	6.50	6.65

Note the lack of response in grain and sugar yields to added P at soil P Index 3

Yields of winter wheat and sugar beet on soils at different K Indices with and without freshly applied potash

		K Index				
		0	1	2-	2+	3
		mg K/L				
Wheat		59	90	150	210	320
Grain yield t/ha	No added K	8.00	10.80	11.35	11.40	11.40
	100 kg K_2O added	9.50	11.00	11.35	11.40	11.40
Sugar beet		90	105	150	210	
Sugar yield t/ha	No added K	5.00	6.00	7.00	7.60	
	310 kg K_2O added	6.65	7.00	7.65	8.05	

Note the lack of response in grain yield to added K at soil K Indices 2-, 2+ and 3

June 2010

Section 1: Principles of nutrient management and fertiliser use

Maintaining the soil Index

The amount of phosphate and potash required for maintenance, in kg P_2O_5/ha and kg K_2O/ha, can be calculated knowing the yield of the crop that was removed from the field and its nutrient content. Typical values for the content of phosphate and potash in crops are given in Appendix 5. For cereals only grain yields are required, if the straw is baled and removed the offtake in grain plus straw is given based on the average offtake in straw determined in experiments.

The phosphate and potash recommendations in this *Manual* for arable crops grown on Index 2 soils are based on typical yields as indicated in the individual crop recommendations (for example, 8 t/ha for winter wheat). For arable crops other than potatoes, where expected yield is significantly greater or less than typical, the amounts of phosphate and potash likely to be removed by the crop should be calculated as shown below. This amount should be used as the maintenance application at the target Index. For potatoes, only potash should be adjusted in this way.

Example

Winter barley was grown on a soil at P Index 2 and K Index 2- and the yield of grain was 8 t/ha. The straw was baled and removed.

Because the soil P and K Indices are 2 and 2- respectively, the fertiliser application should be calculated so that the nutrients removed in the harvested crop are replaced. From Appendix 5, winter barley (straw removed) contains 8.4 kg of P_2O_5 and 10.4 kg K_2O per tonne of grain yield. The following applications should maintain the existing soil Index levels.

Phosphate: 8 (t grain) multiplied by 8.4 (kg P_2O_5/t grain) = 67 kg P_2O_5/ha

Potash; 8 (t grain) multiplied by 10.4 (kg K_2O/t grain) = 83 kg K_2O /ha

It is good practice to apply the maintenance applications regularly. As noted above, if this is done on an annual basis the freshly applied phosphate and potash fertilisers add phosphate and potash for the next crop to be sown. However, if a large amount of either fertiliser is justified because a responsive crop is to be grown then it is acceptable to apply a reduced amount or none for the following crop.

The amount of phosphate removed by potatoes usually is much less than the amount applied. The residue from this surplus application can be taken into account for the following crop.

If replacing the phosphate and potash removed in the harvested crop does not maintain the appropriate Index, it is because there is a slow transfer of phosphate and potash into a less-readily available pool. For this reason it is important to sample the soil and have it analysed every 3-5 years to check that the target Index is being maintained.

No replacement application is shown for magnesium for a number of reasons. The amounts of magnesium removed in a harvested crop tend to be small, perhaps 10-15 kg MgO/ha and it appears that this amount of magnesium can be released during the weathering of clay minerals in many soils. Consequently the amount of exchangeable magnesium in soil tends to change only slowly. Rather than suggest an annual replacement, it is better to monitor change in exchangeable magnesium and when this declines to Mg Index 1 consider applying magnesium especially to sensitive crops like potatoes, sugar beet and some vegetable crops.

Section 1: Principles of nutrient management and fertiliser use

Building up or running down soil Indices

To raise the level of a soil at Index 0 and 1 to Index 2 requires the application of more fertiliser than that needed for the maintenance dressing. The amount of extra fertiliser to apply will be decided on the basis of the costs involved (see below). Organic manures are a useful source of phosphate, potash and magnesium for increasing soil Indices. Soil analysis every third year may be worthwhile where a build up approach is adopted to ensure that the desired levels are not greatly exceeded.

Large amounts of phosphate and potash may be required to raise the crop-available phosphate and potash in the soil by one Index, and it is difficult to give accurate amounts. However, as an example, to increase soil phosphate by 10 mg P/litre may need 400 kg P_2O_5/ha as a phosphate fertiliser (i.e. 850 kg/ha of triple superphosphate). To increase soil potash by 50 mg K/litre may need 300 kg K_2O/ha as a potash fertiliser (i.e. 500 kg/ha of muriate of potash). To apply such large amounts in one dressing is expensive. Consequently, smaller amounts of phosphate and potash are shown as the build up applications in the Table below and in the Tables in Sections 4 to 8. These amounts have been calculated using the mid-point values for each Index (as mg/litre) compared to the mid-point value of Index 2, the target Index. Using these amounts **in addition to the maintenance dressing** should result in the soil Index increasing by one level over 10-15 years where arable crops and grass are grown and 5-10 years where vegetables are grown frequently.

	Arable and forage crops, grassland		Vegetables	
	Phosphate (P_2O_5)	Potash (K_2O)	Phosphate (P_2O_5)	Potash (K_2O)
	kg/ha			
Index 0	60	60	150	150
Index 1	30	30	100	100

<u>Note</u>. In some situations (e.g. phosphate for potatoes), the crop is likely to respond to larger amounts of nutrients than would be recommended using the above adjustments. The recommendations given in the Tables in Sections 4 to 8 are the higher of:

- The rate of nutrient required for maximum crop response.
- The rate of nutrient based on the maintenance dressing plus the amount suggested to increase the soil Index according in the above table.

Where a more rapid increase in the soil Index level is required, larger amounts of fertiliser may be applied but they should be ploughed-in and well mixed with the soil by cultivation. In this situation frequent soil analysis is recommended to ensure that the desired Index level is not exceeded and it should be noted that the analytical values may fall as the equilibrium between the crop-available and less readily available pools of each nutrient is established.

Where the soil Index is well above the target level, then the fertiliser approach could allow for a gradual decline to the target level. For many arable crops and grassland it could be appropriate not to apply any fertiliser rather than a smaller amount unless the crop, like some vegetables, responds to a small amount placed near the seed. If plant available phosphate and potash are being run down then it is essential to follow the decline by regular soil analysis to ensure that the level does not fall below the appropriate target Index.

June 2010

Section 1: Principles of nutrient management and fertiliser use

Examples of adjusting the application of phosphate and potash according to the soil Index and yield

Adjusting the application of phosphate, kg P_2O_5/ha, rounded to the nearest 5 kg

	P Index				
	0	1	2[a]	3	4
Winter wheat, straw incorporated					
10 t/ha	140	110	80	0	0
8 t/ha (standard yield for recommendations)	120	90	60	0	0
6 t/ha	105	75	45	0	0

Adjusting the application of potash, kg K_2O/ha, rounded to the nearest 5 kg

	K Index					
	0	1	2-[a]	2+	3	4
Spring barley, straw removed						
8 t/ha	185	125	90	60	0	0
6 t/ha (standard yield for recommendations)	130	100	70	40	0	0
4 t/ha	110	80	50	20	0	0

a The maintenance application, to replace the phosphate and potash removed in the yield of harvested crop shown in the table, on Index 2 soils. The amount is calculated as shown in the illustration above using the data in Appendix 5. For responsive crops (e.g. potatoes), a higher rate will be needed to ensure there is no limitation to yield.

- The replacement application shown at P Index 2 and K Index 2- is increased on Index 1 and 0 soils. At Index 0 and 1 a crop response to applied phosphate and potash is possible, but the yield is likely to be less than that on a soil at the target Index. The extra amount of phosphate and potash should achieve a long-term build up in the soil Index.
- Where the soil Index is above the target, fertiliser rates may be reduced or omitted for one or several years until the soil Index falls to the target. This can result in significant financial savings.

Section 1: Principles of nutrient management and fertiliser use

Regular soil analysis is essential to ensure the Index does not fall below the target.

Building up crop-available levels of phosphate and potash in soil, or letting them decline, is a decision for the farmer to make. However, two facts should be remembered.

- Experimental evidence shows that applying the maintenance dressing of phosphate and potash fertilisers to soils at Index 0 and 1, even with the suggested addition for build up, will rarely increase the yield of many arable crops to that achieved on Index 2 soils given the maintenance dressing alone.
- Continuing to apply phosphate and potash as fertilisers or manures to soils at P and K Index 4 and above is an unnecessary expense. Also in the case of phosphate, when soils over enriched with phosphate are eroded into inland surface water there is a high risk that the resulting eutrophication will adversely affect the quality of the water.

Soil P Index and loss of phosphate from soil

Particular care should be taken to avoid building up crop-available phosphate above the target Index. To do so is an unnecessary cost and recent research has shown that there is an increased risk of phosphate transfer from soil to surface water when soils are at P Index 3 and above. Much of this transfer is on eroded soil. To reduce the risk of phosphate transfer every effort should be made to minimise the chance of soil erosion and prevent the unnecessary build up of available soil phosphate. *The CoGAP* advises that application of phosphate fertilisers should not exceed the amounts recommended in this *Manual*, which are sufficient for economic crop production. For fields at soil P Index 3 or above, care should be taken to avoid the total combined input of phosphate from organic manure and fertiliser exceeding the total amount of phosphate removed by a crop. This will help avoid raising the soil P level above that necessary for economic crop production. Phosphate transferred from soil to surface fresh water bodies may cause algal blooms and other adverse effects on the biological balance in the water.

Potash use in sandy and sandy loam soils

Sandy and sandy loam soils together with other soils containing very little clay, have a limited capacity to hold potash. On such soils it is almost impossible to achieve the appropriate soil K Index level shown above. For sandy loams it is generally possible to maintain soil at 150 mg K/litre (Index 2-) but for sands and loamy sands, the realistic upper limit is 100 mg K/litre (upper Index 1). Adding potash fertilisers to try to exceed these values will result in movement of potash into the subsoil where it may only be available to deep-rooted crops. On sands, it is preferable to apply and cultivate into the topsoil an amount of potash fertiliser each year to meet the potash requirements of the crop to be grown.

Section 1: Principles of nutrient management and fertiliser use

Responsive situations

Crops such as potatoes, some field vegetables and forage maize may respond to fresh applications of phosphate even where the soil P Index is at or slightly above the target Index level for the rotation. For these crops, special methods of application such as placement or band spreading may be recommended.

For some crops, particularly small seeded vegetables, starter applications of phosphate placed close to the seed improve early germination and establishment even on soils at P Index 3 and above. The amount of phosphate applied as a starter dose together with the amount added in the base dressing should not exceed the amount of phosphate required to replace that removed by the previous crop on soils at P Index 2 and above. **Caution: there is a serious risk of damage to germinating seedlings if base fertiliser formulations containing potash salts are used for starter fertilisers.**

Magnesium

Soil analysis gives the quantity of readily available magnesium in mg Mg/litre of soil, along with an Index (see Appendix 3). The analysis is done on the same soil extract as that used to determine potassium. Potatoes and sugar beet are susceptible to magnesium deficiency and may show yield responses to magnesium fertiliser on soils at Mg Index 0 and 1. Other arable crops may show deficiency symptoms at soil Mg Index 0 but seldom give a yield response to applications of magnesium. Deficiency symptoms often occur early in the growing season when root growth is restricted, for example by soil compaction or excessive soil moisture, but they disappear as the roots grow and thoroughly explore the soil for nutrients. Soil should be sampled and tested regularly, every 3–5 years. For soils at Mg Index 0, 50 to 100 kg MgO/ha can be applied every three or four years.

Where the Mg Index is low and soil acidity needs to be corrected, applying magnesian limestone may be cost-effective. An application of 5 t/ha of magnesian limestone will add at least 750 kg MgO/ha, and this magnesium will become crop-available over many years. However, if used too frequently, the soil Mg Index can exceed 3. In this situation care should be taken to ensure that there is sufficient available potash in soil to ensure that there is no risk of potash deficiency in the crop being grown.

Magnesium recommendations are given as kg MgO/ha not kg Mg/ha. Conversion tables are given in Appendix 8.

Section 1: Principles of nutrient management and fertiliser use

Sulphur

(Section 8 contains additional information for grassland)

Sulphur is an important plant nutrient and plants need about the same amount of sulphur as of phosphorus. Historically, the crop's requirement for sulphur has been met from fertilisers that contain sulphur (e.g., ammonium sulphate) or contained sulphur as a co-product (e.g., calcium sulphate is a co-product in single superphosphate but not in triple superphosphate or ammonium phosphates). Many of these sources containing sulphur are no longer widely used. Additionally, in the past, large amounts of sulphur were released into the atmosphere from industrial processes and this sulphur was deposited on land. However, atmospheric deposition has declined greatly in recent years and levels in 2007 were only about 10% of those in 1980. Consequently, crops should be monitored for signs of sulphur deficiency.

The map shows sulphur deposition in England, Wales and Northern Ireland in 2006.

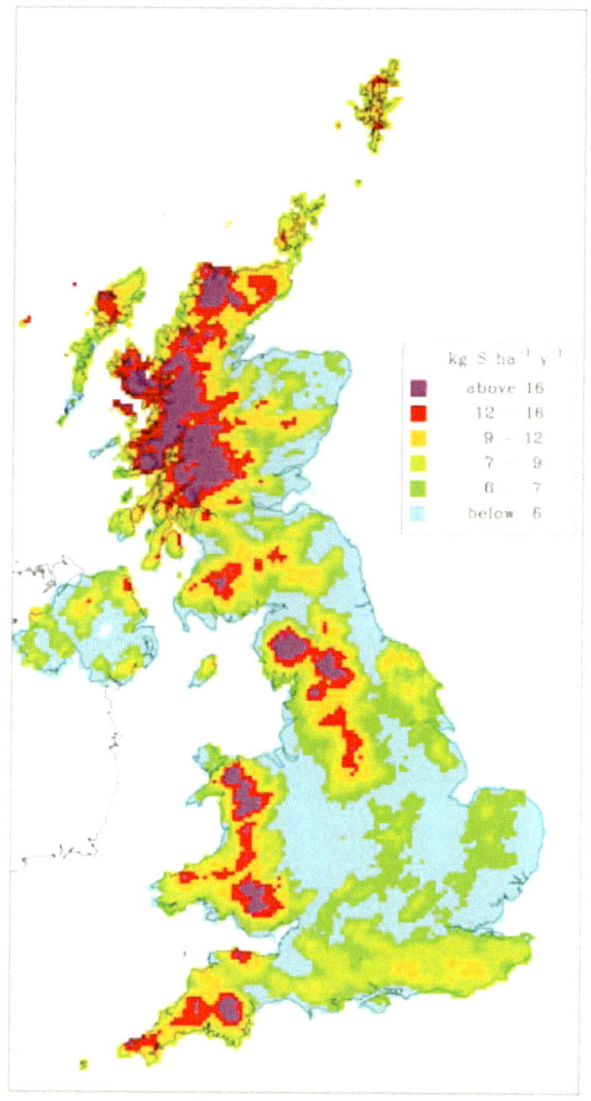

Section 1: Principles of nutrient management and fertiliser use

The occurrence of sulphur deficiency

There is an increasing risk of sulphur deficiency in England and Wales in a wide range of crops including cereals, oilseed rape, brassica vegetables, peas and grass. Oilseed rape and grass grown for silage are particularly sensitive to sulphur deficiency. As atmospheric deposition of sulphur continues to decline, it is likely that the risk of deficiency will affect an increasingly wide range of crops grown on many different soil types.

Currently the best guide for assessing the risk of sulphur deficiency is soil type and field location. Sandy, shallow or medium textured soils that contain little organic matter are most susceptible to sulphur deficiency. Sulphate-S (SO_4^{2-}, the form of sulphur taken up by crops) is not retained in soil because it is soluble in water and is easily leached. Deep silty or clay soils, or fields that have received regular applications of organic manure and organic soils are less likely to show deficiency. Sulphur is retained in soil in organic matter and can become available to plants, like nitrogen, when organic matter is decomposed by soil microbes.

Diagnostic methods

Sulphur deficiency causes paling of young leaves and crop stunting that can easily be confused with nitrogen deficiency (which usually affects older leaves first). In oilseed rape, middle and upper leaves can show interveinal yellowing, and flower petals are unusually pale.

Leaf analysis is a useful guide to diagnosing deficiency in cereals, oilseed rape and grass but interpretative criteria have not yet been established for other crops. Although analytical results may be available too late to correct deficiency in the current crop, they can be useful for decisions on sulphur use for future crops. The procedures for plant sampling and interpretation of analytical results are given with each crop recommendation table. The malate/sulphate ratio test, developed for oilseed rape and cereals, can give a prediction of likely deficiency.

Soil analysis for sulphate-S to 90 cm depth can help identify severely deficient soils but in other situations is not as reliable a guide as leaf analysis.

Sulphur recommendations are given as kg SO_3/ha not kg S/ha. Conversion tables are given in Appendix 8.

Sodium

Sodium is recommended for certain crops (e.g. sugar beet, carrots). Soil analysis can be used to identify the need to apply sodium for sugar beet. Sampling should be carried out according to the procedure described in Appendix 3. The extraction procedure is the same as that used for potassium and magnesium and is described in *Specification for topsoil and requirements for use. British Standard BS 3882: 2007*. The results are available together with those for potassium and magnesium. The interpretation of soil analysis results is given in the crop recommendation tables. At recommended rates, sodium should have no adverse affect on the physical condition of soils.

Section 1: Principles of nutrient management and fertiliser use

In grassland systems, it is important to maintain an adequate amount of sodium in livestock diets. Sodium application will not have any effect on grass growth but has been reported to improve the palatability of grass.

Sodium recommendations are given as kg Na_2O/ha not kg Na/ha. Conversion tables are given in Appendix 8.

Micronutrients (trace elements)

Micronutrients or trace elements are those crop nutrients required in small amounts for essential growth processes in plants and animals. Some micronutrients that are essential for animals are not required by plants but the animal usually acquires them via the plant. In practice only a few micronutrients are known to be present in such small amounts in soil in England and Wales that there is a risk of deficiency in plants and animals. Deficiency is most frequently related to soil type, soil pH, soil structural conditions and their effect on root growth, and crop susceptibility.

Visual symptoms of a deficiency of a specific micronutrient can be confused with those produced by other growth problems. Consequently visual diagnosis of a micronutrient deficiency should, where possible, be confirmed by plant and/or soil analysis.

Deficiencies affecting crop growth

Boron (B): Deficiency can affect sugar beet, brassica crops and carrots on light textured soils with a pH above 6.5, particularly in dry seasons. Symptoms include death of the apical growing point and growth of lateral buds. In sugar beet, there is blackening at the leaf base and beneath the crown ('heart-rot'). Carrots can show a darkening of the root surface ('shadow'). Soil analysis prior to growing a susceptible crop is recommended. When extracted with hot water, a value less than 0.8 mg B/litre dry soil is associated with a risk of deficiency. Leaf analysis is also a useful diagnostic guide and a value lower than 20 mg B/kg dry-matter may indicate deficiency (although the deficiency value varies between crop species).

Copper (Cu): Deficiency is not widespread in crops but can occur mainly in cereals on sands, peats, reclaimed heathland and shallow soils over chalk. Sugar beet may also be affected. In cereals, symptoms are yellowing of the tip of the youngest leaf followed by spiralling and distortion of the leaf. Ears can be trapped in the leaf sheath and those that emerge have white tips. Barley awns can become white. Soil analysis is useful for identifying whether deficiency is likely. When extracted with EDTA, a value lower than 1 mg Cu/litre dry soil indicates possible deficiency. The copper content of the leaf does not reliably indicate the copper status of the plant.

Iron (Fe): Deficiency occurs commonly in fruit crops grown on calcareous soils, but is not a problem in annual field crops. Symptoms in fruit are yellowing of the young leaves with veins remaining green. Deficiency cannot be reliably diagnosed using soil or plant analysis.

Section 1: Principles of nutrient management and fertiliser use

Manganese (Mn): Manganese is the micronutrient most commonly deficient in field crops. Deficiency occurs in many crops on peaty, organic and sandy soils at high pH but can occur less severely on other soils when over-limed. Susceptible crops include sugar beet, cereals and peas. In cereals, deficiency often shows as patches of pale green, limp foliage. Sugar beet leaves develop interveinal mottling and leaf margins curl inwards. Dried peas show internal discolouration when the pea is split ('marsh spot'). Leaf analysis provides a reliable means of diagnosis with a value lower than 20 mg/kg dry matter indicating possible deficiency. Soil analysis is not a reliable guide to deficiency.

Molybdenum (Mo): Deficiency is associated with acid soils and is not generally a problem in limed soils. Cauliflower may be affected and symptoms include restricted growth of the leaf lamina ('whiptail'). Soil or plant tissue analyses may be used to diagnose molybdenum deficiency. Soil analysis is usually by extraction with acid ammonium oxalate ('Tamms reagent'). The leaf and curd content, in dry material, is around 2.0 mg Mo/kg in normal cauliflower plants and around 0.35 mg Mo/kg in deficient plants.

Zinc (Zn): Deficiency is rarely found in field crops. In the few cases where deficiency has been found, it has been on sandy soils, with a high pH and phosphate status. Top fruit and forest nursery stock are most likely to be affected. Leaf analysis is the most useful diagnostic guide and, in susceptible crops, less than 15 mg Zn/kg dry matter may indicate deficiency. Soil analysis usually is by EDTA extraction; for susceptible crops, a value less than 0.5 mg Zn/kg indicates a risk of probable deficiency while less than 1.50 mg Zn/kg indicates possible deficiency.

Deficiencies affecting livestock performance

The availability of cobalt, copper and selenium does not restrict grass growth, but too little in grazed crops can led to deficiency in some animals. Where a deficiency has been correctly diagnosed, treatment of the animal with the appropriate trace element is usually the most effective means of control, though application of selenium and cobalt to grazing pastures is effective.

Fertiliser Types and Quality

It is important to select the most appropriate and cost-effective material from the many different types of fertiliser that are available in both solid and fluid forms. The following features of a fertiliser should be considered:

- The total concentration, and the ratio of nutrients in the fertiliser.
- The chemical form of each nutrient.
- The physical quality of a solid fertiliser and its suitability for accurate spreading.
- The form of a liquid fertiliser, true solution or suspension.
- The cost of the nutrients.

Section 1: Principles of nutrient management and fertiliser use

The total concentration, and the ratio of nutrients in the fertiliser

The total concentration of each nutrient in a fertiliser has to be declared. Some fertilisers (straight fertilisers) contain just one nutrient, whereas many fertilisers contain more than one nutrient. The requirements for nutrient declaration are controlled by legislation contained in *EC Regulation No. 2003/2003, The EC Fertilisers (England and Wales) Regulations 2006 (SI 2486)* and in *The Fertilisers Regulations 1991 as amended* (see Section 9). The nutrient content of some common fertilisers is given in Appendix 7.

The concentration of nutrients in fluid fertilisers may be expressed as kg nutrient per tonne of fertiliser product (w/w basis) or as kg nutrient per cubic metre (1000 litres) of fertiliser product (w/v basis). To convert from one basis to the other, it is necessary to know the specific gravity of the fertiliser. The fertiliser supplier can provide this information.

Concentration as w/v (kg/m^3) = concentration as w/w (kg/t) x specific gravity

The concentration of the nutrient or nutrients in a fertiliser dictates the application rate. When there is more than one nutrient the ratio should be reasonably close to the required application of each nutrient.

The chemical form of each nutrient and its availability for crop uptake

Some nutrients can be present in different chemical forms which may differ in their immediate availability for uptake by roots. The requirements of the declaration are controlled by *The EC Fertilisers (England and Wales) Regulations 2006 (SI 2486), EC Regulation 2003/2003* or, in some cases, by *The Fertilisers Regulations 1991 as amended* (see Section 9).

The physical nature and quality of the fertiliser and its suitability for accurate application

Fertilisers may be sold in many different physical forms, some of which may be difficult to apply accurately. Fertilisers of the same type but from different sources can vary significantly in their physical characteristics. Good fertiliser practice must include accurate, uniform application as well as correct decisions on rate and timing. Inaccurate application of fertiliser will result in uneven crops with lower than expected yields and the quality may be poor. Over application may result in adverse environmental impact from pollution. It is important for good spreading of solid fertiliser that the particle size is consistent, free of lumps and low in dust, while the components of a fluid fertiliser should remain as a uniform solution or suspension.

The cost of the nutrients

The cost of the nutrients in fertilisers can vary significantly. When comparing fertiliser prices it is necessary to calculate and compare the cost of each kg of nutrient. It is cost per hectare, not cost per tonne, which determines the economics of a fertiliser application. Low cost fertilisers can have a poor chemical or physical quality. The availability of the nutrients for crop uptake must also be considered as well as the ability of the fertiliser spreader or sprayer that is used to apply the fertiliser accurately.

Section 1: Principles of nutrient management and fertiliser use

Nitrogen fertilisers

Ammonium nitrate (33.5-34.5% N); ammonium sulphate (21% N, 60% SO_3), calcium ammonium nitrate or CAN (26-28% N): The nitrate-N is immediately available for crop uptake, the ammonium-N can be taken up directly but is quickly converted to nitrate by soil microbes.

Urea (46% N): Before uptake by plants, urea-N must first be converted to ammonium-N by the enzyme urease that occurs in all soils. This process usually occurs quickly and does not significantly delay the availability of the nitrogen for crop uptake. Typically, around 20% of the nitrogen content of applied urea may be lost to the atmosphere as ammonia. As a result less nitrogen is available for crop use and emissions may lead to impacts on biodiversity and human health. Losses are more closely related to soil moisture and weather conditions than to soil type, and may be minimised if urea is applied shortly before rain is expected, and/or is shallowly cultivated into the soil. Urea is a low-density material which, in prilled form, can be less easy to spread accurately over wide bout widths when using spinning disc equipment.

Liquid nitrogen (18-30% N): Liquid nitrogen fertilisers are solutions of urea and ammonium nitrate. The nitrogen is in forms that are quickly available for crop uptake. Solutions based on urea alone will contain no more than 18% N because at low ambient temperatures urea crystallises out of solution.

Phosphate fertilisers

Water-soluble phosphate: Ammonium phosphates (diammonium phosphate (DAP) and monoammonium phosphate (MAP)) and superphosphates (mainly triple superphosphate (TSP) but occasionally single superphosphate (SSP)) contain phosphate mainly (93 – 95%) in a water-soluble form.

Water-insoluble phosphate: There are many different types of water-insoluble phosphate with different chemical and physical characteristics. The fertiliser declaration should give details of the amount of phosphate soluble in different acid extractants. This information does not indicate the effectiveness of these sources of phosphate on different soil types. Care should be taken not to compare the solubility of water-insoluble phosphates in different reagents (for example, formic acid and citric acid) that extract different forms of phosphate.

Lack of water-solubility does not mean the phosphate is unavailable to crops. Finely ground, reactive phosphate rocks, for example, with close to zero water-solubility, can be used successfully as a phosphate source on grassland where the surface soil pH is maintained below pH 6.0. Some other water-insoluble phosphates are an effective source of phosphate under appropriate soil and weather conditions, and in these situations seek advice from a FACTS Qualified Advisor about their use.

Potash and magnesium fertilisers

Potash in most common sources of potash fertiliser is quickly available to crops. The most common potash source is muriate of potash (MOP) which is potassium chloride (60% K_2O). Kainit and sylvinite are naturally-occurring mixtures mainly of potassium chloride and sodium chloride. Potassium sulphate (SOP, 50% K_2O) is used for some high-value crops. Potassium

Section 1: Principles of nutrient management and fertiliser use

nitrate supplies both potassium and nitrogen and is used as a source of both nutrients when added to irrigation systems. Some magnesium fertilisers are quickly available (e.g. kieserite (typically 25% MgO) and Epsom salts (16% MgO)) though others are only slowly available (e.g. calcined magnesite and magnesian limestone). Where both lime and magnesium are needed, magnesian limestone will often provide a cheap though slow acting source of magnesium. An application of 5 t/ha of magnesian limestone will provide at least 750 kg MgO/ha and the magnesium become slowly available to crops over time.

Sulphur fertilisers

Ammonium sulphate (60% SO_3) and kieserite (typically 52-55% SO_3) provide sulphur in an immediately available form. Gypsum (calcium sulphate) is somewhat less soluble but is an effective source. Elemental sulphur must be oxidised to sulphate before it becomes available for uptake. The speed of oxidation depends largely on particle size and only very finely divided ('micronised') elemental sulphur will be immediately effective. Where particle size is larger, elemental sulphur fertilisers can be used to raise the sulphur supply capacity of the soil over a longer period of time.

Fertiliser Application

Accurate and even application of fertilisers is very important in order to maximise the benefits from their use to improve crop yield and quality and profitability. Even where correct decisions have been made on the amount of fertiliser to apply, inaccurate application, uneven spreading or spreading into hedgerows or ditches can cause a range of potentially serious problems, including:

- Uneven crops.
- Lodging and disease.
- Reduced yields and poor or uneven crop quality at harvest.
- More risk of the transfer of nutrients to watercourses at field margins causing nutrient pollution.
- More risk of causing botanical changes in hedgerows and field margins.

Spreading fertilisers and organic manures as uniformly and accurately as is practically possible to the cropped area is a requirement in NVZs. Avoiding spreading into the edges of hedgerows and ditches is a requirement of Cross Compliance.

Fertiliser spreaders and sprayers should be regularly maintained and serviced, replacing worn-out parts as necessary. Spreaders should be calibrated for rate of application every spring and whenever the fertiliser type is changed. To do this, follow the manufacturer's instructions.

To check spreading uniformity, catch-trays can be used. Ideally this should be done annually or whenever faulty spreading is suspected. Computerised analysis of the data will give the Coefficient of Variation (CV) which is a measure of the non-uniformity of spreading. Where the CV is larger than 15%, significant inaccuracies in fertiliser spreading are occurring. Action should then be taken to improve the performance of the spreader.

Section 1: Principles of nutrient management and fertiliser use

More information is given in the leaflet *Fertiliser Spreaders – Choosing, Maintaining and Using* (Agricultural Industries Confederation), or the new guidance book: *Spreading Fertilisers and Applying Slug Pellets* (BCPC).

For manure and slurry spreaders, checks should be made for mechanical condition before spreading and for rate of application during spreading.

Protection of the environment

Farmers are encouraged to adopt a systematic approach to fertiliser planning to help minimise losses of nutrients from agricultural production systems which can, directly or indirectly, have adverse effects on water and air quality with wide ranging effects on loss of biodiversity. Recommendations in this *Manual* for the use of fertilisers and manures are based on economic criteria but matching nutrient applications and soil levels as closely as possible to crop requirement will go a long way to help minimise the risk of nutrient loss from soil to water or to air. Practices that make the best economic use of nutrients also help protect the wider environment. These include:

- Regular soil analysis every 3-5 years for P, K and Mg Index and pH.
- Identification of the SNS Index every spring before applying nitrogen fertiliser.
- Estimation or measurement of the nutrient contents of any organic manure applied.
- Taking account of all other sources of nutrients before deciding on fertiliser application rates.
- Where appropriate, soil or plant tissue analysis to help with decisions on application of sulphur or micronutrients.
- Use of a recognised nutrient recommendation system.
- Regular calibration and tray testing of fertiliser spreaders and calibration of manure spreaders.
- Rapid incorporation of organic manures after application to tillage land or use of trailing hose, trailing shoe or injection equipment for slurry.

Some areas of agricultural land are subject to various types of regulation or agreement for maintenance or improvement of the environment and farmers may have to comply with restrictions on the use of liming materials, fertilisers and manures. **Farmers in these areas may have to modify the recommendations in this Manual to comply with any specific requirements.**

Nitrate

The EC Nitrate Directive, adopted in 1991, requires Member States to introduce controls on agriculture in water catchments where the nitrate (NO_3) concentration in waters either exceeds, or is at risk of exceeding, 50 mg/litre or where there is a risk of eutrophication of surface waters. To comply with these requirements, Nitrate Vulnerable Zones (NVZs) have been designated in the UK. Since December 1998, farmers of land within these areas have had to comply with mandatory NVZ rules (*Defra NVZ Guidance Booklets – see Section 9*).

Section 1: Principles of nutrient management and fertiliser use

Within NVZs, there are closed periods during which manures containing more than 30% readily available nitrogen (cattle or pig slurry, poultry manures and liquid digested sludge) must not be applied although limited applications may be made by registered organic farmers. Closed periods also apply to manufactured fertiliser nitrogen though there are some specified exceptions.

Also within NVZs, the average crop-available nitrogen application rate in manufactured fertiliser and manure to certain crop types must not exceed specified (Nmax) limits.

An important requirement of the Action Programme is that the amount of crop-available nitrogen applied in manufactured fertiliser and manure should not exceed crop requirement and that full allowance should be made for nitrogen available from the soil organic matter and previous crop residues.

Incidental losses of nitrogen to waters should be minimised. This type of loss can be related to a specific farm activity such as manure or fertiliser application that has taken place under unsuitable conditions. Fertilisers or manures should not be applied to steeply sloping, frozen hard, snow-covered or water-logged soil or during rain as these conditions greatly increase the risk of run-off.

The risk of loss of nitrate by leaching can be reduced by ensuring that the amounts of nitrogen applied from all sources are no greater than the crop requires and by applying nitrogen in organic manures and fertilisers under suitable conditions close to the time when the nitrogen in them is needed for crop growth.

Getting nitrogen fertiliser applications right is important for reducing the amount of nitrate leached but choice of crop, autumn crop cover, cultivations, organic manure use and grassland management all have a major impact on the quantity of nitrate lost. Accurate records of past fertiliser use and the regular calibration of fertiliser application machinery will increase the accuracy of fertiliser decisions and application to land.

Ammonia

Agriculture is by far the largest source of all ammonia emissions to the atmosphere, mainly from livestock manures but also from nitrogen fertilisers, especially urea. Ammonia is emitted from solid manures and slurries in livestock housing, stores and following manure spreading to land. The greatest losses are from livestock housing and spreading manures.

Following emission, ammonia can be deposited onto land or water, either nearby or after being transported in the atmosphere over considerable distances. Emission of ammonia can be damaging to the environment because it:

- Contributes to acidification of the soil.
- Adds nitrogen to habitats (e.g. heathland) that are damaged by nutrient inputs.
- Is a precursor to fine particles of ammonium sulphate which have impacts on human health.

Section 1: Principles of nutrient management and fertiliser use

There are strong pressures within the EU to reduce ammonia emissions from agriculture. Large pig and poultry farms are already covered by permits (*Environmental Permitting (England and Wales) Regulations 2007*, SI 3538), which implement the EU Integrated Pollution Prevention and Control (IPPC) Directive. There may be controls on a wider range of farms in the future.

The most effective means of reducing ammonia emissions are to:

- Incorporate solid manure and slurry into the soil on tillage land, preferably by ploughing soon after broadcast spreading.

- Apply slurries with an injector or band-spreader (trailing hose or trailing shoe) (see Section 2).

These 'low emission' spreading techniques reduce ammonia emissions typically by 30-70% compared to conventional 'broadcast' spreading. They are more cost-effective than modifying housing or stores. Where control measures are implemented for housing or stores (e.g. fitting a store cover or roof), it is especially important to use low emission spreading techniques to prevent the conserved ammonia from being lost when the manure is applied to land.

Where possible, urea should not be applied under conditions that promote ammonia loss: warm, drying soils and calcareous soils.

Adopting these techniques to conserve ammonia from late winter and spring applications of manures and slurries will also improve the supply of available nitrogen for crop uptake and will decrease the need for nitrogen fertilisers.

Nitrous oxide

Nitrous oxide (N_2O) is a potent greenhouse gas that is emitted naturally by all soils. Greenhouse gases are associated with climate change and there is strong pressure to limit their emission. The concentration of nitrous oxide in the atmosphere has increased over the past century and part of this increase is attributed to agriculture. Nitrous oxide is formed in soil during nitrification (conversion of ammonium-N to nitrate-N) and denitrification (conversion of nitrate to nitrogen) and its formation tends to be greatest where soil aeration allows both of these processes to occur. This tends to be at water contents around field capacity. Emission of nitrous oxide from the soil tends to increase following the addition of crop residues, manures, fertilisers or any form of mineralisable organic matter that will produce ammonium-N or nitrate-N.

Emission of nitrous oxide can be minimised by ensuring ammonium-N and nitrate-N concentrations in the soil do not exceed crop requirement. Using the amount of nitrogen recommended in this Manual, together with care in fertiliser and manure application, will help minimise both nitrous oxide emission and nitrate leaching.

Section 1: Principles of nutrient management and fertiliser use

Phosphorus

There is increasing environmental concern over the phosphorus status of many inland surface waters and the role of agriculture as a source of phosphorus. Phosphorus is an important nutrient with regard to water quality because small increases in concentration can cause eutrophication (nutrient enrichment) of fresh waters. The effects of eutrophication include algal blooms, fish death, excessive weed growth, poor water clarity and loss of species diversity. These effects are visually unattractive, interfere with water use and ecology, and can be hazardous to animal and human health. The EU *Water Framework Directive* is focussing attention on the need to control eutrophication due to phosphorus movement from soils by requiring all surface waters to have good ecological and chemical status by 2015.

Wastewater from sewage treatment works and agriculture are the main sources of phosphorus in surface waters. Research has shown that the amount of phosphorus transferred from agricultural land can be sufficient to cause eutrophication. Phosphorus moves from soil to water by:

- Surface run-off of recently spread fertilisers and manures
- Erosion of soil particles containing phosphorus
- Particulate and soluble phosphorus in drain outflows

Key measures to reduce the risk of phosphorus movement to water are:

- Following the recommendations in this *Manual* to maintain the target level of crop-available soil phosphate and avoiding any unnecessary build-up above the target Index (see page 37) and taking full account of the phosphate content of organic manures (see Section 2).

- Minimising the risk of soil erosion by following advice contained in *The Code of Good Agricultural Practice* and *Controlling Soil Erosion (MAFF PB 4093)* and guidance in think**soils** from the Environment Agency.

- Avoiding surface applications of all organic manures (solid or liquid) when soils are snow-covered, frozen hard, waterlogged, deeply cracked, or on steeply sloping ground adjacent to watercourses (see *The Code of Good Agricultural Practice*).

- Applying inorganic fertiliser in appropriate amounts as annual dressings rather than as a single, large dressing, except where the aim is to increase the soil Index. Such applications should be ploughed in. If there is a significant risk of surface run-off entering watercourses (e.g. poorly drained clay soils on sloping land) phosphate fertilisers should not be applied when there is a risk of heavy rainfall. This will reduce the risk of phosphorus run-off which can pollute surface waters.

Section 2: Organic manures

	Page
Introduction	56
Livestock manures	56
Allowing for the nutrient content of livestock manures	61
Cattle, pig, sheep, duck or horse solid manures – total and available nutrients	62
Poultry manures – total and available nutrients	64
Cattle slurry and dirty water – total and available nutrients	65
Pig slurry – total and available nutrients	67
Using livestock manures and fertilisers together (with examples)	69
Sewage sludge (biosolids)	74
Biosolids – total and available nutrients	75
Allowing for the nutrient content of sludge (with example)	77
Compost	79
Industrial 'wastes'	82

Section 2: Organic manures

Introduction

Organic manures applied to agricultural land may be produced on the farm (slurries, farmyard manures (FYM) and poultry manures) or supplied from other sources such as treated sewage sludges (commonly called biosolids), composts and industrial 'wastes' such as paper crumble, food industry by-products etc. These materials are valuable sources of most major plant nutrients and organic matter. Careful recycling to land allows their nutrient value to be used for the benefit of crops and soil fertility, which can result in large savings in the use of inorganic fertilisers. In the case of phosphate, such recycling helps conserve the limited global resource of this essential crop and animal nutrient.

Organic manures, particularly solid manures, add useful amounts of organic matter to soils. Their use can improve water holding capacity, drought resistance and structural stability, as well as the biological activity of soils. These improvements are most likely to be achieved where regular manure applications are made. The maintenance and enhancement of soil organic matter levels is a cross compliance requirement of farmers receiving the Single Payment. Care should be taken during application not to cause soil compaction, which may have a detrimental effect on crop growth and increase the risk of surface run-off.

Organic manures can present a considerable environmental risk if not handled carefully. Guidance on avoiding pollution is given in *Protecting our Water, Soil and Air*: *A Code of Good Agricultural Practice for farmers, growers and land managers* (Defra, 2009). In Nitrate Vulnerable Zones (NVZs), the amount of **total nitrogen** in applied organic manures must not exceed 250 kg/ha in any 12 month period at the field level. In some situations, lesser amounts may be appropriate. For example, the amount of **crop available nitrogen** supplied by organic manures should not exceed the amount of nitrogen recommended for the next crop. In some fields, it may be necessary to limit organic manure applications in order to avoid excessive enrichment of soil phosphorus levels. In NVZs, it is mandatory to follow the NVZ Action Programme which includes spreading rate and timing restrictions on the application of organic manures. Guidance booklets are available from Defra (see Section 9).

Livestock Manures

Manure management planning

It is essential to plan the handling and use of manures on a farm. This will ensure that good use is made of the nutrient content ('content' is used rather than the more accurate 'concentration' as this is common usage) of the manures and that the risks of causing environmental pollution are minimised. *The Code of Good Agricultural Practice* (Defra, 2009) provides guidance on how to prepare a Manure Management Plan (see Defra *Manure Management Plan – a step by step guide for farmers*). This plan includes a field risk assessment which will help in deciding when, where and at what rate to apply solid manures, slurry and dirty water, thereby reducing the risks of causing water pollution and transfer of pathogens to water. In NVZs, the preparation of a field risk assessment is a requirement of the Action Programme (NVZ Guidance Leaflet 8).

When planning manure management systems, information is needed on the quantity and nutrient content of livestock manures produced on a farm. This depends on a number of factors, including the number and type of livestock, the diet and feeding system, the volume of dirty water and rainwater entering storage facilities, and the amount of bedding used. Although the

Section 2: Organic manures

volume of manure to be managed will vary considerably with the amount of water introduced into the system (often doubling the volume of slurry to be handled), estimates of the quantities of excreta produced by livestock are useful for calculating manure storage needs, and manure nutrient contents (ex housing and storage) for nutrient planning at the farm level. The table below shows typical outputs of undiluted excreta and nutrient outputs (ex housing and storage) for selected livestock types. In NVZs, it is a mandatory requirement to have 26 weeks storage capacity for pig slurry and poultry manure, and 22 weeks storage capacity for all other livestock slurries (NVZ Guidance Leaflet 4).

Estimated quantities of excreta and nutrient outputs at the end of the housing/storage period.

Type of livestock	Age, liveweight or milk yield	Housing period or occupancy % of year	Output at end of housing/storage period			
			Undiluted excreta (t or m^3)	Nitrogen (N)[d] (kg)	Phosphate (P$_2$O$_5$)[d] (kg)	Potash (K$_2$O)[e] (kg)
Cattle						
Dairy cow	> 9000 litres	60[a]	14.0	69[b]	31	56
	6-9000 litres	60[a]	11.6	60[b]	26	46
	< 6000 litres	60[a]	9.2	46[b]	20	37
Dairy heifer replacement	13 months to first calf	50	7.3	30[b]	12.5	29
Beef suckler cow	> 500 kg	50	8.2	41[b]	15.5	33
	< 500 kg	50	5.8	30[b]	12	23
Beef cattle	13-25 months	50	4.7	25[b]	7.9	19
	3-13 months	50	3.6	17[b]	5.0	15
Pigs						
1 sow + litter	Litter up to 7kg	100	4.0	18	13.5	14
Finishers (dry-meal fed)	7-13 kg	71	0.34	1.0	0.34	1.2
	13-31 kg	82	0.6	4.2	1.8	2.1
	31-66 kg	88	1.2	7.7	3.9	4.2
	66 + kg	86	1.6	10.6	5.6	5.6
Poultry						
1000 Laying hens (caged)	17 weeks and over	97	41	400	350	390
1000 Broilers		85	19[c]	330	220	340
1000 Turkeys (male)		90	53[c]	1230	1020	950
1000 Turkeys (female)		88	39[c]	910	740	690
1000 Ducks		83	30[c]	730	730	230

a. 'housing period' for dairy cows includes allowance for time spent in yards/buildings during milking in the grazing season.
b. N.B. These values should NOT be used for calculating NVZ compliance as they do not include excretal N deposited in the field during grazing.
c. excretal output includes litter, where appropriate.
d. nitrogen and phosphate outputs based on nutrient balance estimates.
e. potash outputs based on estimated undiluted excreta volumes and typical potash content of manures (@10% dry matter content for slurries).

June 2010

Section 2: Organic manures

Nutrient content of manures

For nutrient management planning, it is important to know the nutrient content of manures applied to land. The tables on pages 62-69 give typical values of the total nutrient content of manures based on the analysis of a large number of samples.

Owing to farm-specific feeding and manure handling practices, manures produced at a particular livestock unit may have a nutrient content that is consistently different from the values given in the tables. It is therefore worthwhile having the nutrient content of representative manure samples determined by analysis. Rapid on-farm kits (e.g. Agros, Quantofix) can reliably assess the ammonium-N content of slurries (and it is this value that should be entered into the ammonium-N input box on the MANNER-*NPK* analysis screen), but laboratory analysis is necessary for other nutrients. Laboratory analyses should include dry matter (DM), total N, total P_2O_5, total K_2O, total SO_3, total MgO and ammonium-N (NH_4-N). Additionally, nitrate-N (NO_3-N) should be measured for well composted FYM and poultry manures and slurry that has been treated aerobically, and uric acid-N for poultry manures. Hydrometers can be used to measure slurry dry matter content and, where dry matter varies, to estimate nutrient analyses by adjusting previous laboratory results or the typical values in the cattle (page 65-67) and pig slurry (page 67-69) tables. Appendix 6 provides information on the calculation and interpretation of laboratory manure analysis results.

It is important that sampling is carried out carefully and that representative samples are provided for analysis (see Appendix 6 for guidance on sampling). The optimum sampling frequency will vary depending on how manures are managed on the farm, but at least two samples per year are recommended coinciding with the main spreading periods.

Whether using typical values for the nutrient content of manure, or the results of analysis, the availability of the nutrients for crop uptake must be assessed before the fertiliser replacement value of a manure application can be calculated. Values for the availability of manure nutrients from different application timings and methods have been determined from detailed research studies.

Principles of nitrogen supply and losses

Nitrogen is present in manures in two main forms:

- *Readily available nitrogen* (i.e. ammonium-N as measured by N meters, nitrate-N and uric acid-N) is the nitrogen that is **potentially available** for rapid crop uptake. Slurries and poultry manures are 'high' in readily available-N (typically in the range of 35-70 % of total N) compared with FYM which is 'low' in readily available-N (10-25% of total N) – see diagram below.

- *Organic-N* is the nitrogen contained in organic forms which are broken down slowly to become potentially available for crop uptake over a period of months to years.

Crop available nitrogen is the readily available-N that remains for crop uptake after accounting for any losses of nitrogen. This also includes nitrogen released from organic forms.

Section 2: Organic manures

Following the application of manure to land, there can be losses of nitrogen by two routes. Ammonium-N can be volatilised to the atmosphere as ammonia gas. Following the conversion of ammonium-N to nitrate-N, further losses may occur through nitrate leaching and denitrification of nitrate to nitrous oxide and nitrogen gas under warm and wet soil conditions. To make best use of their nitrogen content, organic manures should be applied at or before times of maximum crop growth – generally during the late winter to summer period.

Ammonia volatilisation

Around 40% of the readily available nitrogen content of manures is often lost following surface application to land. Ammonia loss and odour nuisance can be reduced by ensuring that manures are rapidly incorporated into soils (within 6 hours of application for slurries and 24 hours for solid manures to tillage land). For slurries, shallow injection and band spreading techniques are effective application methods that reduce ammonia emission (typically by 30-70%) compared with broadcast application. Also, slurry band spreading (trailing shoe and trailing hose) and shallow injection application techniques increase the number of spreading days, and cause less sward contamination than surface broadcast applications. These practices will also increase the amount of nitrogen available for crop uptake. Ammonia losses are generally smaller from low dry matter slurries because they more rapidly infiltrate into the soil. Higher dry matter slurries remain on the soil/crop surface for longer leading to greater losses. Losses are also higher when slurries are applied to dry soils under warm weather conditions.

On large pig and poultry units that are permitted under regulations that implement the EU Integrated Pollution Prevention and Control (IPPC) Directive, there is a requirement where solid manures are applied to uncropped land or bare soil to incorporate them into the soil within 24 hours, unless such applications are used to control wind erosion on susceptible soils. Slurry can be applied by an injector, or band spreader, or by any type of splash plate spreader provided slurry is incorporated into the soil within 6 hours of application and provided such equipment is operated to avoid slurry atomisation. Slurry may also be applied by irrigation equipment provided it is applied to a growing crop and such equipment provides a low spreading trajectory (operated at low pressure to create large droplets). If dilute pig slurry (less than 2% dry matter) is applied by irrigation equipment then it does not have to be to a growing crop.

Nitrate leaching

The amount of manure nitrogen leached following land application is mainly related to the soil type, the application rate, the readily available-N content and the amount of rainfall after application. As ammonium-N is rapidly converted in the soil to nitrate-N, manure applications during the autumn or early winter period should be avoided, as there is likely to be sufficient over-winter rainfall to wash a large proportion of this nitrate out of the soil before the crop can use it. Delaying applications until late winter or spring will reduce nitrate leaching and increase the efficiency of utilisation of manure nitrogen. This is particularly important for manures with a high content of readily available-N. In NVZs, there are mandatory closed spreading periods for high readily available-N manures (e.g. slurries, poultry manures), which typically have greater than 30% of their total nitrogen content present as readily available N (see Defra *Guidance for Farmers in Nitrate Vulnerable Zones, Leaflet No. 8 Field application of organic manures*).

Section 2: Organic manures

Typical proportions of different forms of nitrogen in livestock manures

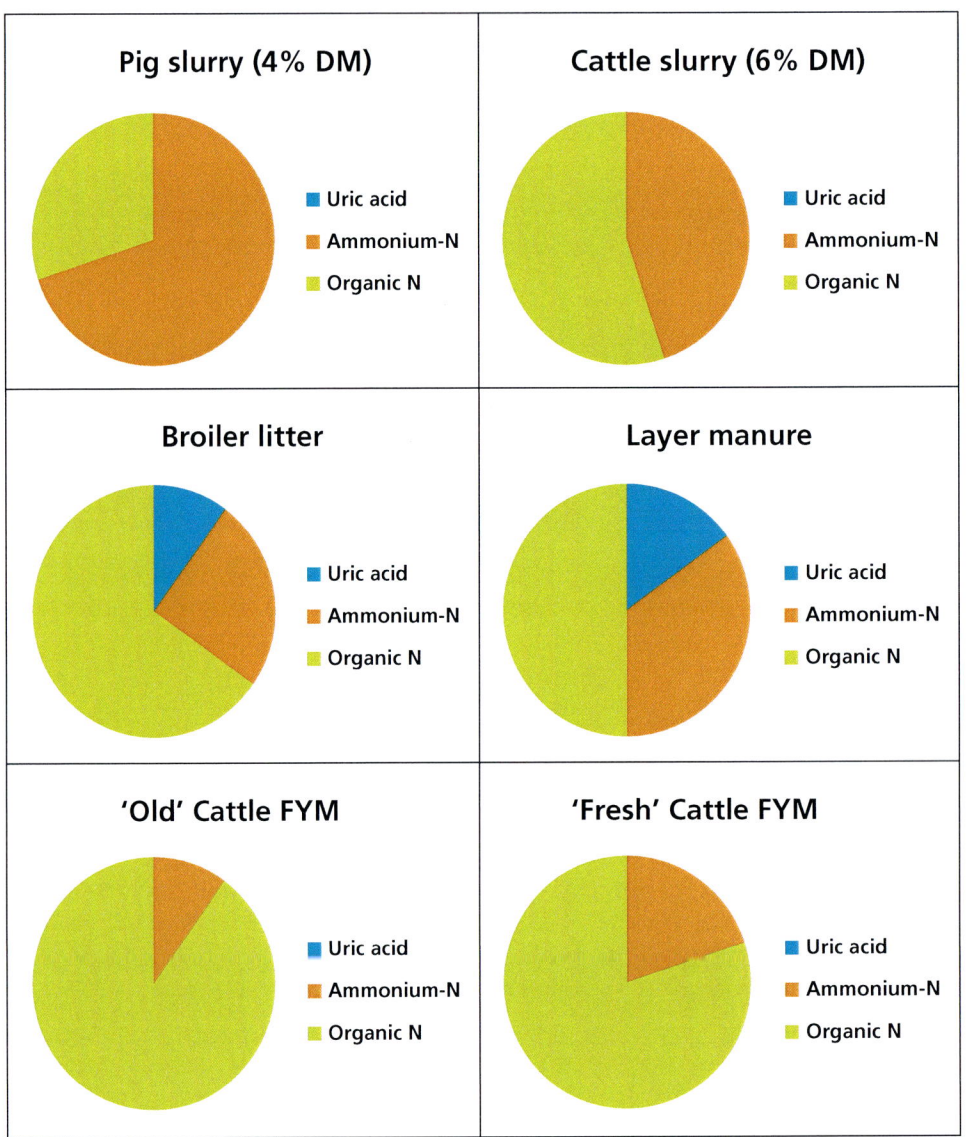

Release of crop-available nitrogen from organic nitrogen

The organic nitrogen content of manures is released (mineralised) slowly over a period of months to years. Where the nitrogen mineralised from the manure is not taken up by the crop in the season following application, nitrate may be lost by leaching during the following over-winter period, or can accumulate in soil organic matter allowing further long-term savings in nitrogen fertiliser inputs. Around 5% of the total nitrogen content of manures may become available for the second crop following application.

Section 2: Organic manures

Allowing for the nutrient content of livestock manures

Nitrogen

Estimates of the percentage of the total nitrogen content that is available for uptake by the next crop from different manure applications are given in the tables on page 63 (FYM), page 64 (poultry manures), page 66 (cattle slurries and dirty water) and page 68 (pig slurries). The tables take account of differences in readily available-N and dry matter content, the effects of application method and timing, soil type and autumn/winter rainfall following application. The footnotes should be used to adjust the values where appropriate.

Where more detailed field specific guidance on the fertiliser nitrogen value of manures is required, use of the MANNER-*NPK* (MANure Nutrient Evaluation Routine) or PLANET decision support systems is recommended. MANNER-*NPK*/PLANET will predict the fertiliser nitrogen value of field applied manures, taking into account the manure type, manure analysis data (total N, ammonium-N, nitrate-N and uric acid-N), soil type, application timing and technique, ammonia-N, nitrate-N and denitrification losses, and the mineralisation of organic-N.

MANNER-NPK decision support system

Where there is uncertainty about the level of residual nitrogen present in the soil, for example, where manures have been applied regularly or at unknown application rates, soil sampling to measure soil mineral nitrogen (SMN) is recommended (see page 26).

Phosphate, potash, magnesium and sulphur

Manures are valuable sources of other nutrients as well as nitrogen, although not all of the total nutrient content is available for the next crop. Typical values for the total and available phosphate and potash contents of farm manures are given in the tables between pages 63 and 69. Nutrients which are not immediately available will mostly become available over a period of years and will usually be accounted for when soil analysis is carried out. The availability of manure phosphate to the next crop grown (50-60%) is lower than from water-soluble phosphate fertilisers. However, around 90% of manure potash is readily available for crop uptake.

Section 2: Organic manures

Where crop responses to phosphate or potash are expected (e.g. soil Indices 0 or 1 for combinable crops and grassland) or where responsive crops are grown (e.g. potatoes or vegetables), the available phosphate and potash content of the manure should be used when calculating the nutrient contribution. Soils at Index 0 will particularly benefit from manure applications. Where the soil is at target Index or above (usually Index 2 or above, see page 37) for phosphate or potash, the total phosphate and potash content of the manure should be used in nutrient balance sheet calculations. For most arable crops, typical manure application rates can supply the phosphate and potash requirement. At soil P Index 3 or above, take care to ensure that total phosphate inputs do not exceed the amounts removed in crops during the rotation. This will avoid the soil P Index reaching an unnecessarily high level. It is important to manage manure applications to supply phosphate and potash **for the crop rotation**.

Manures also supply useful quantities of sulphur and magnesium, but there is only limited data on availability for the next crop grown. Sulphur and magnesium inputs from manures should largely be regarded as contributing to the maintenance of soil reserves.

Cattle, pig, sheep, duck or horse farmyard manure – total and available nutrients

Nitrogen – typical total and readily available nitrogen contents (fresh weight basis)

	Dry matter (%)	Total nitrogen (kg N/t)	'Old' FYM ex – storage[a] Readily available nitrogen (kg N/t)	'Fresh' FYM[b] Readily available nitrogen (kg N/t)
Cattle farmyard manure	25	6.0	0.6	1.2
Pig farmyard manure	25	7.0	1.0	1.8
Sheep farmyard manure	25	7.0	0.7	1.4
Duck farmyard manure	25	6.5	1.0	1.6
Horse farmyard manure	30	7.0	ND	ND

ND = no data.

a. FYM that has been stored for 3 months or more.
b. FYM that is spread straight from the building.

For cattle FYM from organic farms, dry matter and total nitrogen contents are the same as in the table above. However, the readily available nitrogen content of FYM (ex – storage) is 5% of the total, i.e. 0.3 kg/t. For further information see Booklet 4 of the *Managing Livestock Manures* series.

Duck farmyard manure is included here because the availability of its nitrogen is generally lower than that of other poultry manures.

To convert kg/t to units/ton, multiply by 2.

Section 2: Organic manures

Nitrogen – percentage of total nitrogen available to next crop

	Autumn[a] (Aug-Oct, 450 mm rainfall to end March)		Winter[a] (Nov-Jan, 250 mm rainfall to end March)		Spring[a] (Feb-Apr)	Summer[a] use on grassland
	Sandy/ shallow[b]	Medium/ heavy[b]	Sandy/ shallow[b]	Medium/ heavy[b]	All soils	All soils
Surface applied (i.e. not soil incorporated)						
FYM (old and fresh)	5	10	10	10	10	10
Soil incorporated 24 hours after application[d]						
FYM						
- old	5	10	10	10	10	N/A
- fresh [c]	5	10	10	10	15	N/A

N/A = Not applicable

a. The nitrogen availability estimates assume 450 mm of rainfall (after autumn application) and 250 mm (after winter application) up to the end of soil drainage (end March). Where rainfall differs from these amounts, intermediate values of nitrogen availability should be used. For spring or summer applications, rainfall is not likely to cause movement of agronomically important amounts of nitrogen to below crop rooting depth.

b. Sandy/shallow = light sand soils and shallow soils (see Appendix 1) Medium/heavy = medium, deep fertile silt and deep clay soils. Use this category for organic and peaty soils.

c. Fresh FYM – manure which has not been stored prior to land application and has an estimated ammonium-N content of 20% (cattle and sheep FYM) or 25% (pig and duck FYM) of total N. Old FYM – manure which has been stored for 3 months or more and has an estimated ammonium-N and nitrate-N content of 10% (cattle and sheep FYM) or 15% (pig and duck FYM) of the total N.

d. The values assume incorporation by ploughing. Cultivation using discs or tines is less effective in minimising ammonia losses and intermediate values of nitrogen availability should be used.

Phosphate, potash, magnesium and sulphur (fresh weight basis)

	Dry matter (%)	Phosphate			Potash[a]			Total sulphur	Total magnesium
		Total phosphate kg P_2O_5/t	Availability %	Available phosphate kg P_2O_5/t	Total potash kg K_2O/t	Availability %	Available potash kg K_2O/t	kg SO_3/t	kg MgO/t
Cattle farmyard manure	25	3.2	60	1.9	8.0	90	7.2	2.4	1.8
Pig farmyard manure	25	6.0	60	3.6	8.0	90	7.2	3.4	1.8
Sheep farmyard manure	25	3.2	60	1.9	8.0	90	7.2	3.0	1.6
Duck farmyard manure	25	5.5	60	3.3	7.5	90	6.8	2.6	1.2
Horse farmyard manure	30	5.0	60	3.0	6.0	90	5.4	ND	ND

ND = no data.

a. Values of potash may be lower for FYM stored for long periods in the open.

To convert kg/t to units/ton, multiply by 2.

June 2010

Section 2: Organic manures

Poultry manures – total and avilable nutrients

(Information for duck manure is given on page 62)

Nitrogen – typical total and readily available nitrogen content (fresh weight basis)

	Dry matter (%)	Total nitrogen (kg N/t)	Readily available nitrogen (kg N/t)
Layer manure	35	19	9.5
Broiler/turkey litter	60	30	10.5

To convert kg/t to units/ton, multiply by 2.

Percentage of total nitrogen available to next crop following POULTRY MANURE applications (% of total nitrogen)

	Autumn [a] (Aug-Oct, 450 mm rainfall to end March)		Winter [a] (Nov-Jan, 250 mm rainfall to end March)		Spring [a] (Feb-Apr)	Summer [a] use on grassland
	Sandy/shallow [b]	Medium/heavy [b]	Sandy/shallow [b]	Medium/heavy [b]	All soils	All soils
Surface applied (i.e. not soil incorporated)						
Layer manure	10 [15]	25 [30]	25	25	35	35
Broiler/turkey litter	10 [15]	25 [30]	20	25	30	30
Soil incorporated 24 hours after application[c]						
Layer manure	10 [15]	35 [40]	25	40	50	N/A
Broiler/turkey litter	10 [15]	30 [35]	20	30	40	N/A

[use for grassland and winter oilseed rape cropping]

N/A = Not applicable

a. The nitrogen availability estimates assume 450 mm of rainfall (after autumn application) and 250 mm (after winter application) up to the end of soil drainage (end March). Where rainfall differs from these amounts, intermediate values of nitrogen availability should be used. For spring or summer applications, rainfall is not likely to cause movement of agronomically important amounts of nitrogen to below crop rooting depth.

b. Sandy/shallow = light sand soils and shallow soils (see Appendix 1)

Medium/heavy = medium, deep fertile silt and deep clay soils. Use this category for organic and peaty soils.

c. The values assume incorporation by ploughing. Cultivation using discs or tines is less effective in minimising ammonia losses and intermediate values of nitrogen availability should be used.

Section 2: Organic manures

Phosphate, potash, magnesium and sulphur (fresh weight basis)

	Dry matter (%)	Phosphate			Potash			Total sulphur	Total magnesium
		Total phosphate kg P_2O_5/t	Availability %	Available phosphate kg P_2O_5/t	Total potash kg K_2O/t	Availability %	Available potash kg K_2O/t	kg SO_3/t	kg MgO/t
Layer manure	35	14	60	8.4	9.5	90	8.6	4.0	2.6
Broiler/turkey litter	60	25	60	15	18	90	16.2	8.0	4.4

To convert kg/t to units/ton, multiply by 2.

Cattle slurry and dirty water – total and available nutrients

Nitrogen – total and readily available nutrient content (fresh weight basis)

	Dry matter (%)	Total nitrogen (kg N/m³ or /t)	Readily available nitrogen (kg N/m³ or /t)
Slurries/liquids			
Cattle	2	1.6	0.9
	6[a]	2.6[a]	1.2[a]
	10	3.6	1.3
Dirty water	0.5	0.5	0.3
Separated cattle slurries (liquid portion)			
Strainer box	1.5	1.5	0.8
Weeping wall	3	2.0	1.0
Mechanical separator	4	3.0	1.5
Separated cattle slurry (solid portion)	20	4.0	1.0

a. Typical dry matter and nitrogen contents of cattle slurry are shown in bold.

To convert kg/m3 to units/1000 gallons, multiply by 9.

For further information on the total nitrogen and readily available nitrogen of cattle slurry from organic farms, see Booklet 4 of the *Managing Livestock Manures* series.

June 2010

Section 2: Organic manures

Percentage of total nitrogen available to next crop following CATTLE SLURRY and DIRTY WATER applications (% of total nitrogen)

	Autumn [a] (Aug-Oct, 450 mm rainfall to end March)		Winter [a] (Nov-Jan, 250 mm rainfall to end March)		Spring [a] (Feb-Apr)	Summer [a] use on grassland
	Sandy/ shallow [b]	Medium/ heavy [b]	Sandy/ shallow [b]	Medium/ heavy [b]	All soils	All soils
Cattle slurry – liquid Surface applied (i.e. not soil incorporated)						
- 2% DM	5 [10]	30 [35]	30	30	45	35
- 6% DM	**5 [10]**	**25 [30]**	**25**	**25**	**35**	**25**
- 10% DM	5 [10]	20 [25]	20	20	25	20
Soil incorporated 6 hours after application [c]						
- 2% DM	5 [10]	35 [40]	25	35	50	N/A
- 6% DM	**5 [10]**	**30 [35]**	**20**	**30**	**40**	**N/A**
- 10% DM	5 [10]	25 [30]	15	25	30	N/A
Band spread						
- 2% DM	5 [10]	30 [35]	30	30	50	40
- 6% DM	**5 [10]**	**25 [30]**	**25**	**25**	**40**	**30**
- 10% DM	5 [10]	20 [25]	20	20	30	25
Shallow injected						
- 2% DM	5 [10]	30 [35]	35	35	55	45
- 6% DM	**5 [10]**	**25 [30]**	**30**	**30**	**45**	**35**
- 10% DM	5 [10]	20 [25]	25	25	35	30
Dirty water (surface applied)	10 [15]	35 [40]	35	35	50	30
Separated cattle slurry – solid portion Surface applied (i.e. not soil incorporated)	5	10	10	10	10	10
Soil incorporated 24 hours after application [d]	5	10	10	10	15	N/A

[use for grassland and winter oilseed rape cropping]

N/A – not applicable.

a. The nitrogen availability estimates assume 450 mm of rainfall (after autumn application) and 250 mm (after winter application) up to the end of soil drainage (end March). Where rainfall differs from these amounts, intermediate values of nitrogen availability should be used. For spring or summer applications, rainfall is not likely to cause movement of agronomically important amounts of nitrogen to below crop rooting depth.

b. Sandy/shallow = light sand soils and shallow soils (see Appendix 1). Medium/heavy = medium, deep fertile silt and deep clay soils. Use this category for organic and peaty soils.

c. The values assume incorporation by ploughing. Cultivation using discs or tines is less effective in minimising ammonia. Where slurry has been applied in spring or summer and incorporated more quickly than 6 hours or has been deep injected, nitrogen availability will be similar to the shallow injected values.

d. The values assume incorporation by ploughing. Cultivation using discs or tines is less effective in minimising ammonia losses and intermediate values of nitrogen availability should be used.

Section 2: Organic manures

For separated cattle slurry (liquid portion), use the values for 2% dry matter slurry.

Phosphate, potash, magnesium and sulphur (fresh weight basis)

	Dry matter (%)	Phosphate			Potash			Total sulphur	Total magnesium
		Total phosphate (kg P_2O_5/m³ or /t)	Availability %	Available phosphate (kg P_2O_5/m³ or /t)	Total potash kg K_2O/m³ or /t	Availability %	Available potash kg K_2O/m³ or /t	kg SO_3/m³	kg MgO/m³
Slurries/liquids									
Cattle	2	0.6	50	0.3	2.4	90	2.2	0.3	0.2
	6	**1.2**	**50**	**0.6**	**3.2**	**90**	**2.9**	**0.7**	**0.6**
	10	1.8	50	0.9	4.0	90	3.6	1.0	0.9
Dirty water	0.5	0.1	50	0.05	1.0	100	1.0	0.1	0.1
Separated cattle slurries (liquid portion)									
Strainer box	1.5	0.3	50	0.15	2.2	90	2.0	ND	ND
Weeping wall	3	0.5	50	0.25	3.0	90	2.7	ND	ND
Mechanical separator	4	1.2	50	0.6	3.5	90	3.2	ND	ND
Separated cattle slurry (solid portion)	20	2.0	50	1.0	4.0	90	3.6	ND	ND

ND = No data

For further information on the major nutrient content of cattle slurry from organic farms, see Booklet 4 of *Managing Livestock Manures* series (see Section 9).

To convert kg/m³ to units/1000 gallons, multiply by 9.

Pig Slurry – Total and available nutrients

Nitrogen – Typical total and readily available nitrogen content (fresh weight basis)

	Dry matter (%)	Total nitrogen (kg N/m³ or /t)	Readily available nitrogen (kg N/m³ or /t)
Pig slurry – liquid	2	3.0	2.2
	4[a]	3.6[a]	2.5[a]
	6	4.4	2.8
Separated pig slurry (liquid portion)	3	3.6	2.2
Separated pig slurry (solid portion)	20	5.0	1.3

a. Typical dry matter and nitrogen contents of pig slurry are shown in bold.

To convert kg/m³ to units/1000 gallons, multiply by 9.

June 2010

Section 2: Organic manures

Percentage of total nitrogen available to next crop following PIG SLURRY applications (% of total nitrogen)

	Autumn[a] (Aug-Oct, 450 mm rainfall to end March)		Winter[a] (Nov-Jan, 250 mm rainfall to end March)		Spring[a] (Feb-Apr)	Summer[a] use on grassland
	Sandy/shallow[b]	Medium/heavy[b]	Sandy/shallow[b]	Medium/heavy[b]	All soils	All soils
Pig slurry - liquid Surface applied (i.e. not soil incorporated)						
- 2% DM	10 [15]	35 [40]	40	40	55	55
- 4% DM	10 [15]	30 [35]	35	35	50	50
- 6% DM	10 [15]	25 [30]	30	30	45	45
Soil incorporated 6 hours after application[c]						
- 2% DM	10 [15]	45 [50]	35	50	65	N/A
- 4% DM	10 [15]	40 [45]	30	45	60	N/A
- 6% DM	10 [15]	40 [45]	25	40	55	N/A
Band spread						
- 2% DM	10 [15]	35 [40]	40	40	60	60
- 4% DM	10 [15]	35 [40]	35	35	55	55
- 6% DM	10 [15]	30 [35]	35	35	50	50
Shallow injected						
- 2% DM	10 [15]	40 [45]	45	45	65	65
- 4% DM	10 [15]	35 [40]	40	40	60	60
- 6% DM	10 [15]	30 [40]	40	40	55	55
Separated pig slurry - solid portion Surface applied (i.e. not soil incorporated)	5	10	10	10	10	10
Soil incorporated 24 hours after application[d]	5	10	10	10	15	N/A

[use for grassland and winter oilseed rape cropping]

N/A = Not applicable

a. The nitrogen availability estimates assume 450 mm of rainfall (after autumn application) and 250 mm (after winter application) up to the end of soil drainage (end March). Where rainfall differs from these amounts, intermediate values of nitrogen availability should be used. For spring or summer applications, rainfall is not likely to cause movement of agronomically important amounts of nitrogen to below crop rooting depth.

b. Sandy/shallow = light sand soils and shallow soils (see Appendix 1)

Medium/heavy = medium, deep fertile silt and deep clay soils. Use this category for organic and peaty soils.

c. The values assume incorporation by ploughing. Cultivation using discs or tines is less effective in minimising ammonia losses. Where slurry has been applied in spring or summer and incorporated more quickly than 6 hours or has been deep injected, nitrogen availability will be similar to the shallow injected values.

d. The values assume incorporation by ploughing. Cultivation using discs or tines is less effective in minimising ammonia losses and intermediate values of nitrogen availability should be used.

Section 2: Organic manures

For separated pig slurry (liquid portion), use the values for 2% dry matter slurry.

Phosphate, potash, magnesium and sulphur (fresh weight basis)

	Dry matter (%)	Phosphate			Potash			Total sulphur	Total magnesium
		Total phosphate (kg P_2O_5/m^3 or /t)	Availability %	Available phosphate (kg P_2O_5/m^3 or /t)	Total potash kg K_2O/m^3 or /t	Availability %	Available potash kg K_2O/m^3 or /t	kg SO_3/m^3 or /t	kg MgO/m^3 or /t
Pig slurry – liquid	2	1.0	50	0.5	2.0	90	1.8	0.7	0.4
	4	**1.8**	**50**	**0.9**	**2.4**	**90**	**2.2**	**1.0**	**0.7**
	6	2.6	50	1.3	2.8	90	2.5	1.2	1.0
Separated pig slurry (liquid portion)	3	1.6	50	0.8	2.4	90	2.2	ND	ND
Separated pig slurry (solid portion)	20	4.6	50	2.3	2.2	90	2.0	ND	ND

ND = no data

To convert kg/m^3 to units/1000 gallons, multiply by 9.

a. Typical dry matter and nitrogen contents of pig slurry are shown in bold.

Using farm manures and fertilisers together

A planned and integrated manure and inorganic fertiliser policy aims to utilise as much as possible of the nutrient content of manures. Failure to adequately allow for these nutrients, particularly nitrogen, not only wastes money because of unnecessary fertiliser use but can reduce crop yields and quality – e.g. lodging in cereals, poor fermentation in grass silage and low sugar levels in beet.

1. Calculate the quantity of crop available nutrients (equivalent to fertiliser) supplied by each manure application.

2. Identify the fields that are available and that will benefit most from the application of manure. This will need to take account of the accessibility and likely soil conditions in individual fields at the time of application, and the application equipment that is available. Crops with a high nitrogen demand should be targeted first. Fields at low soil P or K Indices will benefit more than those at high Indices.

3. Plan the application rate for each field **ensuring that no more than 250 kg/ha total organic manure nitrogen is applied in any 12 month period** – it is mandatory in NVZs not to exceed this amount. Also, take account of the phosphate content of manures over the *crop rotation* to avoid excessive enrichment of soil phosphorus levels. Make sure that the plans adhere to the NVZ rules in England and Wales. As far as possible, apply manures in the late winter to summer period – this will make best use of the nitrogen content.

Section 2: Organic manures

4. Aim for the manure application to supply no more than 50-60% of the total nitrogen requirement of the crop, with inorganic fertiliser used to make up the difference. This approach will minimise the potential impact of variations in manure nitrogen supply on crop yields and quality.

5. Make sure that manure application equipment is well maintained and suitable for applying the manure in the most effective way, minimising losses of ammonia-N and soil or crop damage. The equipment should be routinely calibrated for the type of manure being applied, using the guidelines contained in Booklet 3 of the *Managing Livestock Manures* series.

6. Following application, use the tables in this section or the MANNER-*NPK* /PLANET decision support systems to calculate the amount of crop available nitrogen supplied from each manure application in each field. Also, use the tables in this book to calculate the amounts of phosphate, potash, sulphur and magnesium applied.

7. Calculate the nutrient requirement of the crop, then deduct the nutrients supplied from manures. This will give the balance that needs to be supplied as inorganic fertiliser (see examples 1 and 2 on pages 72 and 73).

Practical aspects of manure use

- Manures are commonly applied to arable stubbles in the autumn prior to drilling winter cereals. But to make best use of manure nitrogen and to minimise nitrate leaching losses, manures should, if possible, be applied in the late winter to summer period. Band spreaders and other equipment are now available that allow accurate slurry top-dressing across full tramline widths, without causing crop damage. An additional benefit of band spreading is that ammonia emissions are reduced by 30-40% compared with conventional 'splash-plate' surface application, along with odour nuisance.

- Manure applications before spring sown crops (e.g. root crops, cereals and oilseed rape) should be made from late winter onwards to minimise nitrate leaching losses, particularly where high readily available N manures are applied. In NVZs, applications of organic manure with a high readily-available nitrogen content (e.g. slurry, poultry manure and liquid digested sludge) can be made from 1st January onwards on sandy/shallow soils and 16th January onwards on all other soils. Rapid soil incorporation on tillage land (e.g. within 6 hours following surface broadcast application) will minimise ammonia losses and increase manure crop available nitrogen supply.

Section 2: Organic manures

- Manure applications to grassland are best made to fields intended for silage or hay production. Cattle slurry and FYM contain large amounts of potash relative to their readily available nitrogen and total phosphate contents, and are ideally suited to this situation. Solid manure application rates should be carefully controlled to avoid the risk of sward damage and contamination of conserved grass with manure solids, which can adversely affect silage quality. To encourage a low pH and good fermentation, grass cuts following solid manure or late slurry applications should be wilted before ensiling, or an effective silage additive used. To make best use of slurry nitrogen, applications should be made in the late winter to spring period. Slurry applications in summer are likely to be less efficiently utilised because of higher ammonia losses. The use of band spreading and/or shallow injection (5-7 cm deep) techniques will reduce ammonia-N losses (typically by 30-70% compared with surface broadcast application) and herbage contamination.

- Where slurry and solid manure applications are made to grazed grassland, the pasture should not be grazed for at least 4 weeks following application, or until all visible signs of slurry solids have disappeared. This will minimise the risk of transferring disease to grazing livestock. Also, take care to ensure that the manure potash supply does not increase the risk of grass staggers (hypomagnesaemia) in stock through reduced herbage magnesium levels.

- Forage crops, particularly forage maize prior to drilling, provide an opportunity to apply manures in late spring. Manure application rates should be carefully controlled and where possible the manure should be rapidly incorporated into the soil to minimise ammonia-N emissions and odour nuisance. Care should be taken to ensure that such applications do not lead to very high contents of crop-available phosphate in the soil.

- Where manure applications are made before "ready to eat crops" i.e. crops that are generally not cooked before eating, relevant industry guidance should be followed to minimise the risks of pathogen transfer.

Section 2: Organic manures

Example 1.

30 m³/ha of cattle slurry (6% dry matter) is broadcast in early spring before first-cut silage. The soil is at P Index 2 and K Index 2-. Where the slurry is surface broadcast in the spring, allowing for manure nutrients saves up to £95/ha. This potential saving will be less following autumn or winter application, or where soil P or K Indices are above maintenance levels.

	Nitrogen (N)	Phosphate (P_2O_5)	Potash (K_2O)	Financial saving (£/ha)
1. Estimated total nutrients in slurry (kg/m³)	2.6	1.2	3.2	
Analysis of representative sample or typical values from table on pages 65 and 67.				
2. Estimated available nutrients in slurry (kg/m³)				
Nitrogen (table on page 66)				
Phosphate and potash (table on page 67)	0.9[a]	0.6	2.9	
3. Nutrients supplied by slurry that are equivalent to inorganic fertiliser (kg/ha)				
30m³/ha supplies 78 kg/ha total N and 27 kg/ha crop available N	27	36[b]	96[b]	
Potential saving from manure use				£95/ha[d]
4. Nutrient requirements for first-cut silage to produce a yield of 10 t/ha (kg/ha)				
Section 8 of this book	120	40[c]	80[c]	
5. Inorganic fertiliser needed for the silage crop (kg/ha)				
Stage 4 minus Stage 3	93	4	NIL	
Actual saving for next crop from manure use				£85/ha
6. Surplus manure nutrients for subsequent crops that are equivalent to inorganic fertiliser (kg/ha)				
Stage 3 minus Stage 4		NIL	16	
Saving for subsequent crops from manure use				£10/ha

a. Nitrogen availability is 35% of total N (see table on page 66)
b. Total phosphate and potash content used in calculations to maintain soil Indices.
c. Nutrients required for spring application (soil P 2 and K Index 2-)
d. Saving for next crop plus value of surplus manure phosphate and potash which will contribute to the nutrient requirement of future crops.

Assumed fertiliser costs: Nitrogen 60p/kg; phosphate 60p/kg; potash 60p/kg

Section 2: Organic manures

Example 2.

35 t/ha of pig FYM is applied in autumn to a clay soil before drilling winter wheat (8 t/ha grain yield, straw baled). It is NOT rapidly incorporated. The soil is at P Index 2 and K Index 2-. Where the FYM is surface applied in the autumn, allowing for manure nutrients saves up to £309/ha. This potential saving will be less where soil P or K Indices are above maintenance levels.

Stage and calculation procedure	Nitrogen (N)	Phosphate (P_2O_5)	Potash (K_2O)	Financial savings (£/ha)
1. Estimated total nutrients in FYM (kg/t)				
Analysis of representative sample or typical values from table on pages 62 and 63	7.0	6.0	8.0	
2. Estimated available nutrients in FYM (kg/t)				
Nitrogen (table on page 63)				
Phosphate and potash (table on page 63)	0.7[a]	3.6	7.2	
3. Nutrients supplied by FYM that are equivalent to inorganic fertiliser (kg/ha)				
35t/ha supplies 245 kg/ha total N and 25 kg/ha crop available N	25	210[b]	280[b]	
Potential saving from manure use				£309/ha[d]
4. Nutrient requirements for winter wheat (kg/ha)				
Section 4 of this book	220	70[c]	85[c]	
5. Inorganic fertiliser needed for the wheat crop (kg/ha)				
Stage 4 minus Stage 3	195	NIL	NIL	
Actual saving for next crop from manure use				£108/ha
6. Surplus manure nutrients for subsequent crops that are equivalent to inorganic fertiliser (kg/ha)				
Stage 3 minus Stage 4		140	195	
Saving for subsequent crops from manure use				£201/ha

a. Nitrogen availability is 10% of total N (see table on page 63)
b. Total phosphate and potash content used in calculations to maintain soil Indices.
c. Nutrients required for maintenance of soil reserves (soil P Index 2 and K Index 2-)
d. Saving for next crop plus value of surplus manure phosphate and potash which will contribute to the nutrient requirement of future crops.

Assumed fertiliser costs: Nitrogen 60p/kg; phosphate 60p/kg; potash 60p/kg

Section 2: Organic manures

Manure application

It is important that manures are applied evenly and at known application rates. The most important causes of uneven application on farms are the incorrect setting of bout widths and poor attention to machinery maintenance. For both slurries and solid manures, the evenness of spreading is usually better with rear discharge spreaders than side discharge machines. Top-dressing slurry to arable crops in spring can be carried out using tankers or umbilical systems, with boom applicators (fitted with nozzles or trailing-hoses) operating from tramlines. The aim should be to apply all manure types evenly with a coefficient of variation of less than 25%. This is achievable with many commonly used types of manure application equipment provided they are well maintained and calibrated.

Application rates can be calculated simply from a knowledge of the capacity of the slurry tanker or solid manure spreader (by weighing both full and empty machines on a weighbridge), the number of loads applied per field and the field area. An accurate flow meter should be used to measure the slurry application rate of umbilical and irrigation systems. More information about manure spreading systems is in Booklet 3 of the *Managing Livestock Manures* series (see Section 9).

Heavy metals

Livestock manures also contain heavy metals, which on certain soils, for example copper deficient soils, can correct micronutrient (trace element) deficiency. However, in the majority of situations, the accumulation of heavy metals in soil is the more important issue. Pig and poultry manures can contain elevated levels of zinc and copper, which in the long-term (over 100 years), may lead to undesirably high soil levels. Where pig or poultry manures have been applied to land for a number of years and will continue to be applied, it is advisable to sample the soil periodically for analysis for heavy metals.

Sewage sludge (Biosolids)

Treated sludges (commonly called biosolids) are valuable fertilisers and soil conditioners, which have undergone processes to create a product suitable for beneficial use in agriculture.

Where sludges are applied to agricultural land the conditions of *The Safe Sludge Matrix, The Sludge (Use in Agriculture) Regulations and the Code of Practice for Agricultural Use of Sewage Sludge* must be followed. *The Safe Sludge Matrix* provides the minimum standard for sustainable sludge recycling to agricultural land. Where sludges are used on agricultural land, usage must be recorded and the soil tested by the producer. These operating requirements ensure that sludge applications to farmland are strictly controlled, that contents of heavy metals do not accumulate to elevated levels in soils and crops, and that disease risks to humans and livestock are minimised. Before spreading on the field, users are advised to consult their product purchasers concerning any possible use restrictions imposed by the food supply chain.

Nutrient content of sludge

Biosolids are a valuable source of major plant nutrients and organic matter, which can be used by growers to meet crop nutrient requirements and to maintain soil fertility (see the example on

Section 2: Organic manures

page 78). Solid sludges (e.g. digested cake, lime stabilised cake) are the most common products applied to farmland.

Based on the analysis of a large number of samples, typical nutrient content data for the main biosolids types applied to farmland are summarised in the table below. However, the characteristics of these products can vary depending on the individual source and treatment process. Most biosolids products are now supplied by water companies with specific nutrient content data and other information.

Around 50% of the total phosphate content of biosolids is available to the next crop grown, with the remainder becoming available over future years. However, availability may be lower if the biosolids have been tertiary treated using iron and aluminium salts to enhance the removal of phosphorus from wastewater. The phosphate supplied by a biosolids application should be considered over the whole crop rotation by managing inputs in relation to crop offtake and soil analysis. Biosolids contain only small amounts of potash. Useful quantities of sulphur and magnesium are also applied which will help to meet crop needs and contribute to the maintenance of soil reserves. Lime-stabilised sludges also have value as liming materials (neutralising value typically in the range of 2-6% per tonne fresh weight) that can balance the acidifying effects of atmospheric inputs and of inorganic (urea and ammonium-based) nitrogen fertilisers.

Nitrogen supply from sludge

The same factors that affect the supply and losses of nitrogen from livestock manures (i.e. organic nitrogen mineralisation and ammonia volatilisation, nitrate leaching and denitrification) also apply to biosolids. The table on page 76 gives the percentage of biosolids total nitrogen content that will be available for the next crop grown, in relation to product type, timing and application method, soil type and rainfall. Nitrogen will also be supplied to crops in the seasons following biosolids application. In the second year, digested cake has been shown to supply around 10% of the total N applied, and around 5% in the third year.

BIOSOLIDS – total and available nutrients

Nitrogen – Typical total and readily available nitrogen content (fresh weight basis)

	Dry matter (%)	Total nitrogen (kg N/m^3)	Readily available nitrogen (kg N/m^3)
Digested liquid	4	2.0	0.8
		(kg N/t)	(kg N/t)
Digested cake	25	11	1.6
Thermally dried	95	40	2.0
Lime stabilised	40	8.5	0.9
Composted	60	11	0.6

To convert kg/m^3 to units/1000 gallons, multiply by 9.
To convert kg/t to units/ton, multiply by 2.

Section 2: Organic manures

Percentage of total nitrogen available to next crop following biosolids applications (% of total nitrogen)

	Autumn [a] (Aug-Oct, 450 mm rainfall to end March)		Winter [a] (Nov-Jan, 250 mm rainfall to end March)		Spring [a] (Feb-Apr)	Summer [a] use on grassland
	Sandy/ shallow [b]	Medium/ heavy [b]	Sandy/ shallow [b]	Medium/ heavy [b]	All soils	All soils
Surface applied (i.e. not soil incorporated)						
Digested liquid	10 [15]	30 [35]	30	30	40	30
Digested cake	10	15	15	15	15	15
Thermally dried	10	15	15	15	15	15
Lime stabilised	10	15	15	15	15	15
Composted	10	15	15	15	15	15
Soil incorporated after application - 6 hours for liquids and 24 hours for solids [c]						
Digested liquid	10 [15]	35 [40]	25	35	45	N/A
Digested cake	10	15	15	15	20	N/A
Thermally dried	10	15	15	15	20	N/A
Lime stabilised	10	15	15	15	20	N/A
Composted	10	15	15	15	15	N/A
Deep injected (25-30 cm)						
Digested liquid	10 [15]	30 [35]	30	30	50	50

[use for grassland and winter oilseed rape cropping]

N/A = Not applicable

a. The nitrogen availability estimates assume 450 mm of rainfall (after autumn application) and 250 mm (after winter application) up to the end of soil drainage (end March). Where rainfall differs from these amounts, intermediate values of nitrogen availability should be used. For spring or summer applications, rainfall is not likely to cause movement of agronomically important amounts of nitrogen to below crop rooting depth.
b. Sandy/shallow = light sand soils and shallow soils (see Appendix 1)
Medium/heavy = medium, deep fertile silt and deep clay soils. Use this category for organic and peaty soils.
c. The values assume incorporation by ploughing. Cultivation using discs or tines is likely to be less effective in minimising ammonia losses and intermediate values of nitrogen availability should be used.

Section 2: Organic manures

Phosphate, potash, magnesium and sulphur (fresh weight basis)

	Dry matter (%)	Phosphate			Potash			Total sulphur	Total magnesium
		Total phosphate kg P_2O_5/m^3	Availability %	Available phosphate kg P_2O_5/m^3	Total potash kg K_2O/m^3	Availability %	Available potash kg K_2O/m^3	kg SO_3/m^3	kg MgO/m^3
Digested liquid	4	3.0	50	1.5	0.1	90	0.1	1.0	0.3
		kg P_2O_5/t	%	kg P_2O_5/t	kg K_2O/t	%	kg K_2O/t	kg SO_3/t	kg MgO /t
Digested cake	25	18	50	9.0	0.6	90	0.5	6.0	1.6
Thermally dried	95	70	50	35	2.0	90	1.8	23	6.0
Lime stabilised	40	26	50	13	0.8	90	0.7	8.5	2.4
Composted	60	6.0	50	3.0	3.0	90	2.7	2.6	2.0

To convert kg/m^3 to units/1000 gallons, multiply by 9.

To convert kg/t to units/ton, multiply by 2.

Allowing for the nutrient content of biosolids

The same principles apply as have been described on page 61 for livestock manures.

Section 2: Organic manures

Example.

20 t/ha of digested cake is applied in autumn before winter wheat (8 t/ha grain yield, straw baled), grown on a medium soil following a previous cereal crop. The sludge is rapidly incorporated. The soil is at P Index 2 and K Index 2-. Where the digested cake is surface applied in the autumn, allowing for biosolids nutrients saves up to £243/ha. This potential saving will be less where soil P or K Indices are above maintenance levels.

Stage and calculation procedure	Nitrogen (N)	Phosphate (P_2O_5)	Potash (K_2O)	Financial saving (£/ha)
1. Estimate total nutrients in digested cake (kg/t)				
Analysis provided by biosolids supplier or typical values from tables on pages 75 and 77.	11	18	0.6	
2. Estimate available nutrients in digested cake (kg/t)				
Nitrogen (see page 75)				
Phosphate and potash (see page 77)	1.7[a]	9.0	0.5	
3. Nutrients supplied by digested cake that are equivalent to inorganic fertiliser (kg/ha)				
20t/ha supplies 220 kg/ha total nitrogen and 33 kg/ha of crop available N	33	360[b]	12[b]	
Potential saving from biosolids use				£243/ha[d]
4. Nutrient requirements for winter wheat (kg/ha)				
See Section 4 of this book	220	70[c]	85[c]	
5. Inorganic fertiliser needed for the wheat crop (kg/ha)				
Stage 4 minus Stage 3	187	NIL	73	
Actual saving for next crop due to biosolids use				£69/ha
6. Surplus digested cake phosphate for subsequent crops that is equivalent to inorganic fertiliser (kg/ha)	See note e			
Stage 3 minus Stage 4		290	NIL	
Saving for subsequent crops due to biosolids use				£174/ha

a. Nitrogen availability is 15% of total N (see table on page 76)
b. Total phosphate and potash content used in calculations to maintain soil P and K Indices.
c. Nutrients required for maintenance of soil reserves (soil P Index 2 and K Index 2-)
d. Saving for next crop plus value of surplus biosolids phosphate which will contribute to the nutrient requirement of future crops.
e. Some additional nitrogen will be available in the 2nd year (20 kg/ha) and 3rd year (10 kg/ha) after application – see page 75.

Assumed fertiliser costs: Nitrogen 60p/kg; phosphate 60p/kg; potash 60p/kg.

Section 2: Organic manures

Heavy metals

Biosolids products contain heavy metals but at lower concentrations than in the past. On certain soils, for example copper deficient soils, biosolids can correct a trace element deficiency. However in the majority of situations, long-term accumulation of heavy metals in the soil is the more important issue. For biosolids additions, there is a statutory requirement to analyse topsoils for heavy metals before land spreading to ensure that concentrations are below maximum permissible soil levels, and to control annual additions of metals. Limits for soil concentrations and permitted rates of addition of heavy metals are given in Defra *Protecting Our Water, Soil and Air: A Code of Good Agricultural Practice*.

Compost

Composts made from source-separated biodegradable inputs are valuable sources of plant available nutrients and organic matter. They are generally made from landscaping and garden 'wastes' (green compost), and additionally can contain kitchen/catering 'wastes' (green/food compost).

Compost that meets the requirements of the *Quality Compost Protocol* and the associated product standard (BSI PAS 100) is called 'Quality Compost', and may be spread on agricultural land without an *Environmental Permitting Regulations* exemption or a Standard Permit (which other green or green/food composts require). BSI PAS 100 is a specification that requires controls on inputs, the composting process and product quality, and requires the compost producer to establish and implement a quality management system. The Quality Compost Protocol defines the point at which compost can be treated as a product and ceases to be waste. Before spreading on the field, users are advised to consult their produce purchasers concerning any possible use restrictions imposed by the food supply chain.

Where Quality Compost is used in agriculture and field horticulture the following requirements should be met to protect human health and the environment:

- Records of input materials to the composting process.

- Analysis of the Quality Compost giving characteristics and nutrients that will be supplied at a recommended application rate to meet crop needs and legislative requirements.

- Soil analysis results, including heavy metals, prior to field spreading.

Nutrient content of compost

Composts are a valuable source of stable organic matter and crop available nutrients, which can be used by growers to meet crop nutrient requirements and to maintain soil fertility (see the example on page 81). Green compost is the most commonly applied product to land, although increasing amounts of green/food compost will be produced in the next few years.

Section 2: Organic manures

Based on the analysis of a large number of green compost samples, typical nutrient content data are summarised in the table below. Typical analysis data are also summarised for green/food composts, although they are based on more limited sample numbers. The nutrient content of compost products will vary depending on the source materials and treatment process. Most composts are supplied with specific nutrient content data and other relevant information.

Typical total nutrient contents (fresh weight basis)

Compost type	Dry matter (%)	Nitrogen		Total			
		Total (kg N/t)	Readily available (kg N/t)	Phosphate (kg P_2O_5/t)	Potash (kg K_2O/t)	Sulphur (kg SO_3/t)	Magnesium (kg MgO/t)
Green	60	7.5	<0.2	3.0	5.5	2.6	3.4
Green/food	60	11	0.6	3.8	8.0	3.4	3.4

Nutrient supply from compost

The available field experimental data indicate that green compost supplies only very small amounts of crop available nitrogen and that inorganic fertiliser nitrogen application rates should not be changed for the next crop grown. In the case of green/food compost, the available experimental data indicate that around 5% of the total nitrogen applied is available to the next crop grown (irrespective of application timing). Following the repeated use of green and green/food composts long-term soil nitrogen supply will be increased.

As little work has been done on the availability of compost phosphate to crops, it is appropriate to extrapolate from work on livestock manures and sewage sludge which suggests that around 50% of the phosphate will be available to the next crop grown, with the remainder released slowly over the crop rotation. Around 80% of compost potash is in a soluble form and is readily available for crop uptake. Soil analysis will indicate if the phosphate and potash applied in compost is maintaining the target Index for both nutrients. Composts also supply useful quantities of sulphur and magnesium, although there are no data on availability to the next crop grown. Composts also have a small liming value that can balance the acidifying effects of inorganic fertiliser nitrogen additions to soils.

Section 2: Organic manures

Example.

30 t/ha of green compost is applied in autumn to a sandy soil before drilling winter barley (8 t/ha grain yield, straw baled). The soil is at P Index 2 and K Index 2-. Allowing for the green compost nutrient supply saves up to £153/ha. This potential saving will be less where soil P or K Indices are above maintenance levels.

Stage and calculation procedure	Nitrogen (N)	Phosphate (P_2O_5)	Potash (K_2O)	Financial savings (£/ha)
1. Estimated total nutrients in green compost (kg/t)				
Analysis provided by green compost supplier or typical values from table on page 80.	7.5	3.0	5.5	
2. Estimated available nutrients in green compost (kg/t)				
Nitrogen (see page 80)	NIL[a]			
Phosphate and potash (see page 80)		1.5	4.4	
3. Nutrients supplied by green compost that are equivalent to inorganic fertiliser (kg/ha)				
30t/ha supplies 225 kg/ha total nitrogen	NIL	90[b]	165[b]	
Potential saving from green compost use				£153/ha[d]
4. Nutrient requirements for barley (kg/ha)				
See Section 4 of this book	160	70[c]	95[c]	
5. Inorganic fertiliser needed for the barley crop (kg/ha)				
Stage 4 minus Stage 3	160	NIL	NIL	
Actual saving for next crop due to green compost use				£99/ha
6. Surplus compost phosphate and potash for subsequent crops that is equivalent to inorganic fertiliser (kg/ha)	NIL	20	70	
Stage 3 minus Stage 4				£54/ha
Saving for subsequent crops due to green compost use				

a. Nitrogen availability is negligible or, for practical purposes, nil (see text on page 80)
b. Total phosphate and potash content used in calculations to maintain soil P and K Indices.
c. Nutrients required for maintenance of soil reserves (soil P Index 2 and K Index 2-)
d. Saving for next crop plus value of surplus compost phosphate and potash which will contribute to the nutrient requirement of future crops.

Assumed fertiliser costs: Nitrogen 60p/kg; phosphate 60p/kg; potash 60p/kg.

Section 2: Organic manures

Industrial 'wastes'

The recycling of industrial 'wastes' to agricultural land is controlled by the *Environmental Permitting Regulations*. These Regulations allow the spreading of some industrial 'wastes' onto agricultural land under an exemption providing that certain conditions are met i.e. that they can be shown to provide 'agricultural benefit' or 'ecological improvement'. The application of such 'wastes' must be registered with the Environment Agency who will supply advice on the Regulations and their interpretation. The typical nutrient content of selected industrial 'waste' materials that are commonly recycled to farmland are summarised below.

Paper crumble – total nutrients

Typical total nutrient content (fresh weight basis)

	Dry matter (%)	Total nitrogen (kg N/t)	Total phosphate (kg P_2O_5/t)	Total potash (kg K_2O/t)	Total sulphur (kg SO_3/t)	Total magnesium (kg MgO/t)
Chemically/physically treatment	40	2.0	0.4	0.2	0.6	1.4
Biologically treated	30	7.5	3.8	0.4	2.4	1.0

ND = no data
To convert kg/t to units/ton, multiply by 2.

Following the application of chemically/physically treated paper crumble nitrogen 'lock-up' commonly occurs due to the wide carbon: nitrogen ratio of the paper crumble which immobilises soil nitrogen. As a general rule, around 0.8 kg of inorganic nitrogen is required per tonne (fresh weight) of paper crumble applied to compensate for the nitrogen 'lock-up' in the soil. As biologically treated paper crumble has a lower carbon: nitrogen ratio, nitrogen 'lock-up' is not usually experienced following land spreading.

Mushroom compost – total nutrients

Typical total nutrient content (fresh weight basis)

	Dry matter (%)	Total nitrogen (kg N/t)	Total phosphate (kg P_2O_5/t)	Total potash (kg K_2O/t)	Total sulphur (kg SO_3/t)	Total magnesium (kg MgO/t)
Mushroom compost	35	6.0	5.0	9.0	ND	ND

ND = no data
To convert kg/t to units/ton, multiply by 2.

Section 2: Organic manures

Water treatment cake – total nutrients

Typical total nutrient content (fresh weight basis)

	Dry matter (%)	Total nitrogen (kg N/t)	Total phosphate (kg P$_2$O$_5$/t)	Total potash (kg K$_2$O/t)	Total sulphur (kg SO$_3$/t)	Total magnesium (kg MgO/t)
Water treatment cake	25	2.4	3.4	0.4	5.5	0.8

To convert kg/t to units/ton, multiply by 2.

Food industry 'wastes' – total nutrients

Typical total nutrient content (fresh weight basis)

	Dry matter (%)	Total nitrogen (kg N/m^3 or /t)	Total phosphate (kg P$_2$O$_5$/m^3 or /t)	Total potash (kg K$_2$O/m^3 or /t)	Total sulphur (kg SO$_3$/m^3 or /t)	Total magnesium (kg MgO/m^3 or /t)
Dairy	4	1.0	0.8	0.2	ND	ND
Soft drinks	4	0.3	0.2	Trace	ND	ND
Brewing	7	2.0	0.8	0.2	ND	ND
General	5	1.6	0.7	0.2	ND	ND

ND = no data
To convert kg/t to units/ton, multiply by 2.

Typically industrial 'waste' materials are supplied to farmers with specific nutrient content data and advice on how to best manage these materials to the benefit of soils and crops.

Section 3: Using the recommendation tables

	Page
Finding the nitrogen recommendation	86
Field Assessment Method	86
Table A. Soil Nitrogen Supply (SNS) Indices for low rainfall areas	91
Table B. Soil Nitrogen Supply (SNS) Indices for moderate rainfall areas	92
Table C. Soil Nitrogen Supply (SNS) Indices for low high areas	93
Table D. Soil Nitrogen Supply (SNS) Indices following ploughing out of grass leys	94
Measurement Method	95
Finding the phosphate, potash and magnesium recommendations	99
Finding the sulphur and sodium recommendations	101
Selecting the most appropriate fertiliser	102
Calculating the amount of fertiliser to apply	102

(See Section 8 for grassland recommendation tables)

To find the correct recommendation for nitrogen, phosphate, potash and magnesium for a particular crop it is important to read the notes that accompany each table.

For nitrogen, Soil Nitrogen Supply (SNS) Index is determined using The Field Assessment Method or The Measurement Method (see Section 1 for details).

For phosphate, potash and magnesium the P, K and Mg Index is based on soil analysis (see Section 1 for details).

More background underlying the principles of nutrient management, including fertiliser applications, for crops is given in Section 1.

Section 3: Using the recommendation tables

Finding the nitrogen recommendation

Fields vary widely in the amount of nitrogen available to a crop before any fertiliser or manure is applied. This variation must be taken into account to avoid inadequate or excessive applications of nitrogen. The Soil Nitrogen Supply (SNS) system has been developed to assign an Index of 0 to 6 that indicates the likely extent of this background nitrogen supply. Once the Index is identified, it can be used in the recommendation tables to indicate the amount of nitrogen, as fertiliser, manure or a combination of both that typically would need to be applied to ensure optimum yield.

The SNS Index for each field can be arrived at either by the 'Field Assessment Method' using records of soil type, previous cropping and winter rainfall or by the 'Measurement Method' using measurements of soil mineral nitrogen (SMN) plus estimates of nitrogen already in the crop (at the time of soil sampling) plus an estimate of available nitrogen from the mineralisation of soil organic matter and crop debris during the period of active crop growth.

Field Assessment Method

(for the Measurement Method, go to page 95)

The Field Assessment Method does NOT take account of the nitrogen that will become available to a crop from any organic manures applied since harvest of the previous crop. The available nitrogen from manures applied since harvest of the previous crop, or those that will be applied to the current crop, should be calculated separately using the information in Section 2, and deducted from the fertiliser nitrogen application rates shown in the recommendation tables.

There are five essential steps to follow to identify the appropriate SNS Index:

Step 1. Identify the soil type for the field.

Step 2. Identify the previous crop.

Step 3. Select the rainfall range for the field.

Step 4. Identify the provisional SNS Index using the appropriate table.

Step 5. Make any necessary adjustments to the SNS Index.

In detail these steps are:

Step 1. Identify soil type for the field

Careful identification of the soil type in each field is very important. The whole soil profile should be assessed to 1 metre depth for arable crops and to rooting depth for shallow-rooting vegetables. Where the soil varies, and nitrogen is to be applied uniformly, select the soil type that occupies the largest part of the field.

The soil type can be identified using the following flow chart that categorises soils on their ability to supply and retain mineral nitrogen. The initial selection can then be checked using the

Section 3: Using the recommendation tables

soil category table below. Carefully assess the soil organic matter content when deciding if the soil is organic (10% to 20% organic matter for the purposes of this *Manual*) or peaty (more than 20% organic matter). If necessary, seek professional advice on soil type assessments, remembering this will need to be done only once.

Soil category assessment – flow diagram

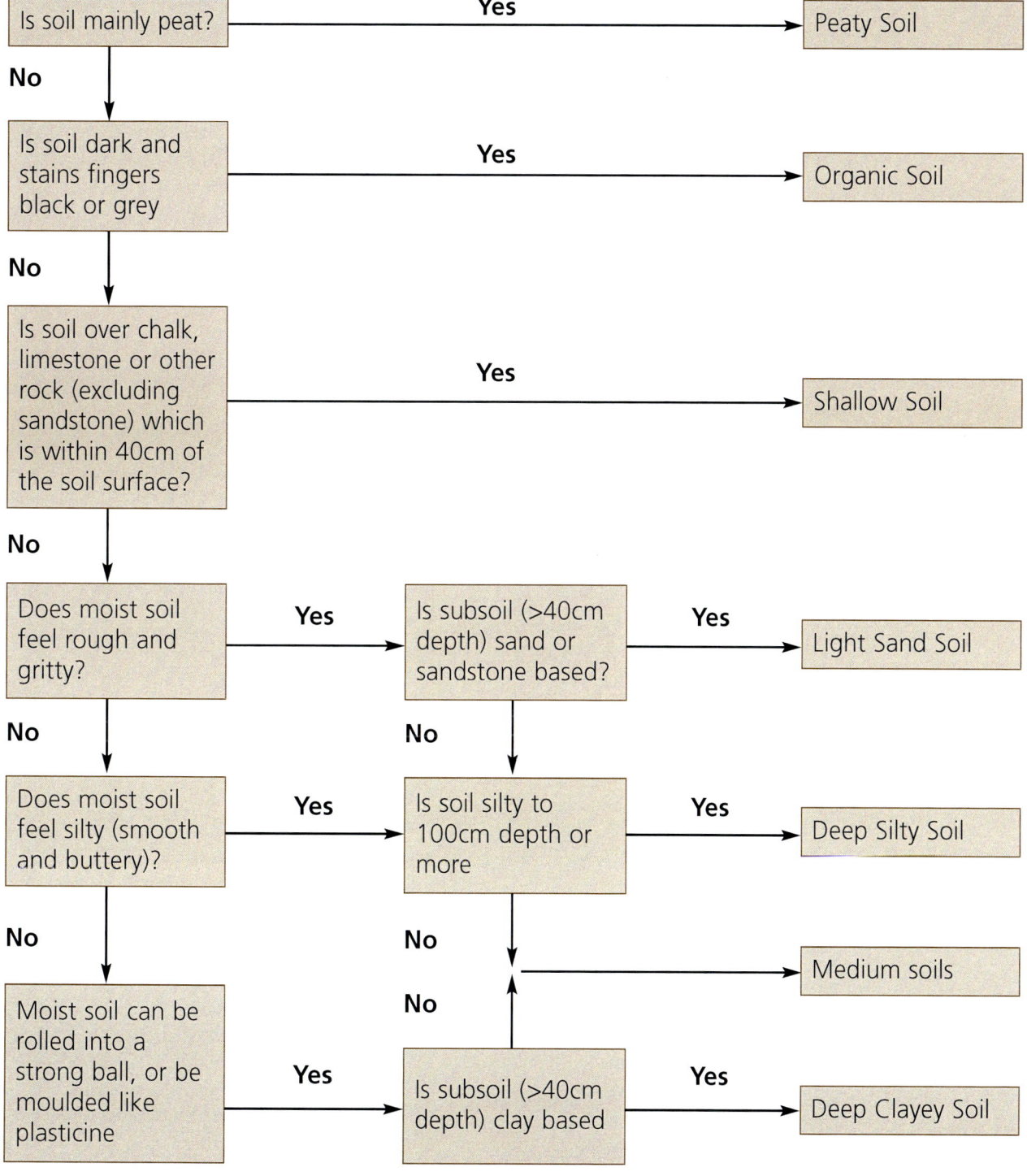

Section 3: Using the recommendation tables

Soil category assessment – table

Soil Category	Description of Soil Types Within Category	Properties
Light sand soils	Soils which are sand, loamy sand or sandy loam to 40 cm depth and are sand or loamy sand between 40 and 80 cm, or over sandstone rock.	Soils in this category have poor water holding capacity and retain little nitrogen.
Shallow soils	Soils over impermeable subsoils and those where the parent rock (chalk, limestone or other rock) is within 40 cm of the soil surface. Sandy soils developed over sandstone rock should be regarded as light sand soils.	Soils in this category are less able to retain or supply nitrogen at depth.
Medium soils	Mostly medium-textured mineral soils that do not fall into any other soil category. This includes sandy loams over clay, deep loams, and silty or clayey topsoils that have sandy or loamy subsoils.	Soils in this category have moderate ability to retain nitrogen and allow average rooting depth.
Deep clayey soils	Soils with predominantly sandy clay loam, silty clay loam, clay loam, sandy clay, silty clay or clay topsoil overlying clay subsoil to more than 40cm depth. Deep clayey soils normally need artificial field drainage.	Soils in this category are able to retain more nitrogen than lighter soils.
Deep silty soils	Soils of sandy silt loam, silt loam or silty clay loam textures to 100 cm depth or more. Silt soils formed on marine alluvium, warp soils (river alluvium) and brickearth soils are in this category. Silty clays of low fertility should be regarded as other mineral soils.	Soils in this category are able to retain more nitrogen than lighter soils and allow rooting to greater depth.
Organic soils	Soils that are predominantly mineral but with between 10 and 20% organic matter to depth. These can be distinguished by darker colouring that stains the fingers black or grey.	Soils in this category are able to retain more nitrogen than lighter soils and have higher nitrogen mineralisation potential.
Peat soils	Soils that contain more than 20% organic matter derived from sedge or similar peat material.	Soils in this category have very high nitrogen mineralisation potential.

Step 2. Identify previous crop

Usually, this is straightforward but sometimes clarification may be needed:

High residual nitrogen vegetables ('High N vegetables') are leafy, nitrogen-rich brassica crops such as calabrese, brussels sprouts and some crops of cauliflower where significant amounts of crop debris are returned to the soil, especially in rotations where an earlier brassica crop has been grown within the previous twelve months. To be available for crop uptake, this organic nitrogen must have had time to mineralise but the nitrate produced must not have been at risk to loss by leaching.

Section 3: Using the recommendation tables

Medium residual nitrogen vegetables ('Medium N vegetables') are crops such as lettuce, leeks and long season brassicas such as Dutch white cabbage where a moderate amount of crop debris is returned to the soil.

Low residual nitrogen vegetables ('Low N vegetables') are crops such as carrots, onions, radish, swedes or turnips where the amount of crop residue is relatively small.

In Table D, 'High N' grassland means average annual applications of more than 250 kg N/ha in fertiliser plus available nitrogen in manure used in the last two years, or clover-rich swards or lucerne.

'Low N' grassland means average annual inputs of less than 250 kg N/ha in fertiliser plus available nitrogen in manure used in the last two years, or swards with little clover.

Annual Rainfall in England and Wales

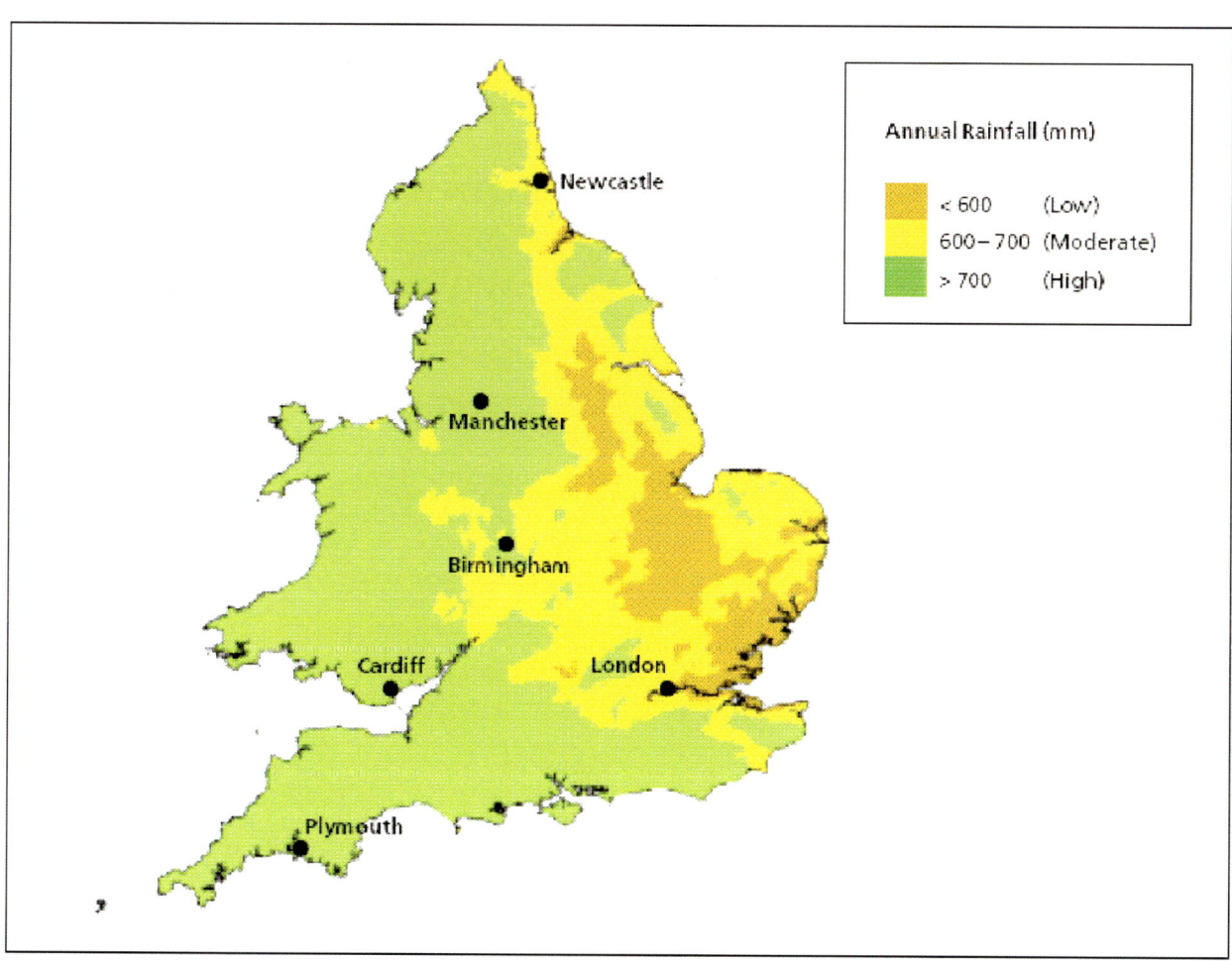

Section 3: Using the recommendation tables

Step 3. Select low, moderate or high rainfall

The appropriate rainfall category should be identified. Ideally an estimate or measurement of excess winter rainfall (EWR, see Section 1) is required because this is closely related to drainage by which nitrate will be lost through leaching. The map above shows long-term average annual rainfall. In unusually wet or dry seasons, it may be appropriate to use actual values of excess winter rainfall and this may result in use of a different SNS Index table. The Met Office has a fee-paying service that can provide information on EWR.

If measurement or estimation of EWR is not possible then the typical annual rainfall can be used. The map on page 89, or preferably local measurement, can be used to decide if the typical annual rainfall is less than 600 mm (usually less than 150 mm excess winter rainfall, 'Low rainfall'), between 600-700 mm (150-250 mm excess winter rainfall, 'Moderate rainfall') or over 700 mm (over 250 mm excess winter rainfall, 'High rainfall'). Using total rainfall rather than EWR is a somewhat less reliable method for deciding on the appropriate rainfall category, especially if an average value rather than a measured one is used.

Step 4. Identify the provisional SNS Index using Table A, B, C or D as appropriate.

Tables A (Low rainfall), B (Moderate rainfall) and C (High rainfall) should be used where the field has not been in grass within the past three years. Select one of these tables according to rainfall for the field. Take account of the footnotes to the tables.

Boxes with a * indicate indices that are not typical but that can occur sometimes in practice. Higher than typical indices can occur where there has been a history of grassland or frequent applications of organic manures. Soil analysis for SMN is recommended in these situations (see page 29).

Boxes that are shaded indicate indices that rarely occur.

If grass has been grown in the previous three years, **also** look at Table D. Select the higher of the Index levels based on the last crop grown (from Table A, B or C) and that based on the grass history (from Tables D1-D4).

Section 3: Using the recommendation tables

Table A.

Soil Nitrogen Supply (SNS) Indices for Low rainfall (500-600 mm annual rainfall, up to 150 mm excess winter rainfall) – based on the last crop grown

	SNS Index						
	0	1	2	3	4	5	6
	SNS (kg N/ha) SNS = SMN (0-90 cm soil depth) + crop N + estimate of net mineralisable N						
	<60	61-80	81-100	101-120	121-160	161-240	Over 240
Light sands or shallow soils over sandstone	Cereals, Low N vegetables, Forage crops (cut)	Sugar beet, Oilseed rape, Potatoes, Peas, Beans, Medium N vegetables, Uncropped land	High N vegetables	*	*	*	*
Medium soils or shallow soils not over sandstone	*	Cereals, Sugar beet, Low N vegetables, Forage crops (cut)	Oilseed rape, Potatoes, Peas, Beans, Uncropped land	Medium N vegetables	High N vegetables [a]	*	*
Deep clayey soils	*	*	Cereals, Sugar beet, Low N vegetables, Forage crops (cut)	Oilseed rape, Potatoes, Peas, Beans, Medium N vegetables[a], Uncropped land	High N vegetables [a]	*	*
Deep silty soils	*	*	Cereals, Sugar beet, Low N vegetables, Forage crops (cut)	Oilseed rape, Potatoes, Peas, Beans, Medium N vegetables[a], Uncropped land	High N vegetables [a]	*	*
Organic soils	All crops – see page 86						
Peat soils	All crops						

a. Index may need to be increased by up to 1 where significantly larger amounts of leafy residues are incorporated (see Step 5). Where there is uncertainty, soil sampling for SMN may be appropriate.

Do not confuse SNS (Soil Nitrogen Supply) and SMN (Soil Mineral Nitrogen).

SMN is the measured amount of mineral nitrogen (nitrate-N plus ammonium-N) in the soil profile.

- SNS = SMN (0-90 cm or maximum rooting depth in shallow soils over rock) + crop N (at time of sampling for SMN) + estimate of available N from mineralisation of organic matter.

Section 3: Using the recommendation tables

Table B.

Soil Nitrogen Supply (SNS) Indices for Moderate rainfall (600-700 mm annual rainfall, or 150-250 mm excess winter rainfall) – based on the last crop grown

	SNS Index						
	0	1	2	3	4	5	6
	SNS (kg N/ha) SNS = SMN (0-90 cm soil depth) + crop N + estimate of net mineralisable N						
	<60	61-80	81-100	101-120	121-160	161-240	Over 240
Light sands or shallow soils over sandstone	Cereals Oilseed rape Potatoes, Sugar beet, Low/Med vegetables Forage (cut)	High N vegetables, Peas, Beans, Uncropped land	*	*	*	*	*
Medium soils or shallow soils not over sandstone	*	Cereals, Sugar beet, Low N vegetables Forage (cut)	Oilseed rape Peas, Beans, Potatoes, Med N vegetables, Uncropped land	High N vegetables	*	*	*
Deep clayey soils	*	Cereals, Sugar beet, Low N vegetables, Forage (cut)	Oilseed rape Potatoes, Peas, Beans, Uncropped land	Med N vegetables	High N vegetables	*	*
Deep silty soils	*	Cereals, Sugar beet, Low N vegetables Forage (cut)	Oilseed rape, Potatoes, Uncropped Land	Peas, Beans, Med N vegetables	High N vegetables	*	*
Organic soils				All crops – see page 86			
Peat soils				All crops			

Do not confuse SNS (Soil Nitrogen Supply) and SMN (Soil Mineral Nitrogen).

- SMN is the measured amount of mineral nitrogen (nitrate-N plus ammonium-N) in the soil profile.
- SNS = SMN (0-90 cm or maximum rooting depth in shallow soils over rock) plus crop N (at time of sampling for SMN) plus an estimate of available N from mineralisation of organic matter.

Section 3: Using the recommendation tables

Table C.

Soil Nitrogen Supply (SNS) Indices for High rainfall (over 700 mm annual rainfall, or over 250 mm excess winter rainfall) – based on the last crop grown

	SNS Index						
	0	1	2	3	4	5	6
	SNS (kg N/ha) SNS = SMN (0-90 cm soil depth) + crop N + estimate of net mineralisable N						
	<60	61-80	81-100	101-120	121-160	161-240	Over 240
Light sands or shallow soils over sandstone	Cereals, Oilseed rape, Potatoes, Sugar beet, Peas, Beans, Low/medium vegetables, Forage crops (cut), Uncropped land	High N vegetables[a]	*	*	*	*	*
Medium soils or shallow soils not over sandstone	*	Cereals, Oilseed rape, Potatoes, Peas, Beans, Sugar beet, Low and medium N vegetables, Forage crops (cut), Uncropped land	High N vegetables	*	*	*	*
Deep clayey soils	*	Cereals, Sugar beet, Oilseed rape, Potatoes, Low and medium N vegetables, Forage crops (cut), Uncropped land	Peas, Beans, High N vegetables	*	*	*	*
Deep silty soils	*	Cereals, Sugar beet, Low N vegetables, Forage crops (cut)	Medium N vegetables, Oilseed rape, Potatoes, Peas, Beans, Uncropped land	High N vegetables	*	*	*
Organic soils					All crops – see page 86		
Peat soils						All crops	

a. Index may need to be lowered by 1 where residues incorporated in the autumn and not followed immediately by an autumn-sown crop.

Do not confuse SNS (Soil Nitrogen Supply) and SMN (Soil Mineral Nitrogen).

- SMN is the measured amount of mineral nitrogen (nitrate-N plus ammonium-N) in the soil profile.

- SNS = SMN (0-90 cm or maximum rooting depth in shallow soils over rock) + crop N (at time of sampling for SMN) + estimate of available N from mineralisation of organic matter.

Section 3: Using the recommendation tables

Table D.

Soil Nitrogen Supply (SNS) Indices following ploughing out of grass leys

The Indices shown in Table D assume that little or no organic manures have been applied. Where silage fields have received the organic manures produced by livestock that have eaten the silage and the manure has been applied in spring, such fields should be regarded as containing nitrogen residues equivalent to a previous grazing history.

See Step 2 of the Field Assessment Method above for definitions of 'High N' and 'Low N' grassland.

	SNS Index		
D1. Light sands or shallow soils over sandstone – all rainfall areas	Year 1	Year 2	Year 3
All leys with 2 or more cuts annually receiving little or no manure 1-2 year leys, Low N 1-2 year leys, 1 or more cuts 3-5 year leys, Low N, 1 or more cuts	0	0	0
1-2 year leys, High N, grazed 3-5 year leys, Low N, grazed 3-5 year leys, High N, 1 cut then grazed	1	2	1
3-5 year leys, High N, grazed	3	2	1
D2. Other mineral soils and shallow soils – not over sandstone – all rainfall areas			
All leys with 2 or more cuts annually receiving little or no manure 1-2 year leys, Low N 1-2 year leys, 1 or more cuts 3-5 year leys, Low N, 1 or more cuts	1	1	1
1-2 year leys, High N, grazed 3-5 year leys, Low N, grazed 3-5 year leys, High N, 1 cut then grazed	2	2	1
3-5 year leys, High N, grazed	3	3	2
D3. Deep clayey soils and deep silty soils in Low rainfall areas (500-600 mm annual rainfall)			
All leys with 2 or more cuts annually receiving little or no manure 1-2 year leys, Low N 1-2 year leys, 1 or more cuts 3-5 year leys, Low N, 1 or more cuts	2	2	2
1-2 year leys, High N, grazed 3-5 year leys, Low N, grazed 3-5 year leys, High N, 1 cut then grazed	3	3	2
3-5 year leys, High N, grazed	5	4	3
D4. Deep clayey soils and deep silty soils in Moderate (600-700 mm annual rainfall) or High (over 700 mm annual rainfall) rainfall areas			
All leys with 2 or more cuts annually receiving little or no manure 1-2 year leys, Low N 1-2 year leys, 1 or more cuts 3-5 year leys, Low N, 1 or more cuts	1	1	1
1-2 year leys, High N, grazed 3-5 year leys, Low N, grazed 3-5 year leys, High N, 1 cut then grazed	3	2	1
3-5 year leys, High N, grazed	4	3	2

Section 3: Using the recommendation tables

Step 5. Make any necessary adjustment to the SNS Index for certain conditions

When using the Field Assessment Method, it is not necessary to estimate the amount of nitrogen taken up by the crop over winter. This is already taken into account in the tables.

Manure history: Where regular applications of organic manures have been made to previous crops in the rotation, increase the Index value by one or two levels depending on manure type, application rate and frequency of application. **The nitrogen contribution from manures applied after harvest of the previous crop should not be considered when deciding the SNS Index. This contribution should be deducted from the recommended nitrogen application rate using the information in Section 2.**

Field vegetables as previous crop: On medium, deep silty or deep clayey soils, nitrogen residues in predominantly vegetable rotations can persist for several years especially in the drier parts of the country. This is likely to be especially evident following 'High or Medium N vegetables'. The SNS tables make some allowance for this long persistency of nitrogen residues but the Index level may need to be adjusted upwards particularly where winter rainfall is low, where the history of vegetable cropping is longer than one year, and in circumstances where larger than average amounts of crop residue or unused fertiliser are left behind (see Footnote to Table A). In rotations where vegetable crops are grown infrequently in essentially arable rotations the Index level may need to be adjusted downwards. Where there is uncertainty, soil sampling for SMN may be appropriate.

In vegetable rotations where a second crop is grown in the summer season, increase the Index for the second crop by one level from that arrived at in Step 4 above if following 'Medium N vegetables', and by one or two levels following 'High N vegetables'. It is important that the growing conditions of the first crop are fully taken into account. For instance, nitrogen may be leached below rooting depth in wet seasons or where excess irrigation has been applied especially on light sandy soils. Analysis for SMN (0-90 cm) before the second crop could be worthwhile.

Fertiliser residues from previous crop: The Index assessments assume that the previous crop grew normally and that it received the recommended rate of nitrogen applied as fertiliser and/or organic manures. The Index should be increased if there is reason to believe nitrogen residues are likely to be greater than normal and these residues will not be lost by leaching. This could occur where a cover crop was sown in autumn and grew well over winter. The Index may need to be adjusted downwards if there is reason to believe nitrogen residues are likely to be smaller than usual.

After any adjustment, the SNS Index can be used in the recommendation tables.

The measurement method

This method is particularly appropriate where the SNS is likely to be large and uncertain (see page 29 and Appendix 2). This includes fields with a history of organic manure application and vegetable rotations where the timing of residue incorporation can strongly affect Soil Mineral Nitrogen (SMN) for the following crop. Nitrogen residues also can be large following outdoor pigs. The SNS Index can be identified using the results of direct measurement of (SMN) to 90 cm depth or maximum rooting depth in shallow soils over rock. The crop nitrogen content (at

Section 3: Using the recommendation tables

the time of soil sampling) and an estimate of net mineralisable nitrogen must be added to the SMN result when calculating the SNS. The Measurement Method is not recommended for peat soils where net mineralisation can be very large and uncertain and the measured SMN may be a relatively small component of SNS. For these soils, the Field Assessment Method or local experience will be better guides to SNS.

Do not confuse SNS (Soil Nitrogen Supply) and SMN (Soil Mineral Nitrogen).

SMN is the measured amount of mineral nitrogen (nitrate-N plus ammonium-N) in the soil profile.

> **SNS = SMN (0-90 cm or maximum rooting depth in shallow soils over rock) + crop N (at time of sampling for SMN) + estimate of available N from mineralisation of organic matter.**

The Measurement Method does NOT take account of the available nitrogen supplied from organic manures applied after the date of soil sampling for SMN. The available nitrogen from manures applied after sampling should be calculated separately using the information in Section 2, and deducted from the nitrogen rate shown in the appropriate recommendation table. **The nitrogen contribution from manures applied before sampling for SMN will be largely taken account of in the measured value and should not be calculated separately**.

When using the Measurement Method there are four steps to follow:

Step 1. Measure SMN

Step 2. Estimate nitrogen already in the crop.

Step 3. Make an allowance for net mineralisable nitrogen

Step 4. Identify SNS Index

In detail these four steps are:

Step 1. Measure SMN

Sampling the soil to 90cm depth is difficult to do manually, especially as a minimum of 15-20 soil cores per field (based on 10 ha) will be needed to obtain a representative sample. Information on sampling and analysis for SMN is in Appendix 2. Note samples should not be frozen but cooled and maintained at less than 5°C until analysed.

Where SMN is measured to only 30 or 45 cm soil depths for shallow rooted crops (for example lettuce, onions and other salad crops), use of the analysis results will underestimate the SNS Index which is based on SMN in the 0-90 cm soil depth. A decision support system such as WELL_N can be used to help interpretation of these soil nitrogen measurements. Alternatively, estimate SMN to 90cm depth by assuming a uniform concentration of mineral nitrogen for all soil layers. Further information is in Section 5 Vegetables and bulbs.

Analysis for SMN is the best method for measuring nitrogen residues following grassland **but is not recommended during the first year after ploughing out**. See page 29 for more details and Appendix 2 for sampling guidelines.

Section 3: Using the recommendation tables

Step 2. Estimate nitrogen already in the crop

Where a crop is present when SMN is measured, the amount of nitrogen already taken up must be estimated. For cereals, this is often a small though important component of the SNS but for oilseed rape, it can be as large as SMN.

The crop nitrogen content in cereals can be assessed according to the number of shoots present (main shoots and tillers), as follows:

Shoot number/m^2	Crop nitrogen content (kg N/ha)
500	5-15
1000	15-30
1500	25-50

Use the smaller crop nitrogen content of the range shown when assessing crops in late autumn and the larger crop nitrogen content for crops in early spring.

In oilseed rape, the crop contains around 50 kg N/ha for every unit of Green Area Index (GAI). Alternatively, the nitrogen content of an average density crop can be assessed by measuring the average crop height.

Crop height (cm)	Crop nitrogen content (kg N/ha)
10	35-45
15	55-65
20	75-85

Add the estimate of nitrogen in the crop to the measured SMN.

Step 3 Make an adjustment for net mineralisable nitrogen

Nitrogen mineralised from soil organic matter and crop debris after soil sampling is a potentially important source of nitrogen for crop uptake. However, in mineral soils of low to average organic matter content (less than about 10%), the amount of net mineralisable nitrogen will be small and for practical purposes, no adjustment is needed when using the recommendations in this *Manual*.

For field vegetable crops allowances are made in the recommendation tables for future mineralisation depending on planting and harvesting dates, see the field vegetables section.

An adjustment may be needed where soil organic matter content is above average or where there has been a history of regular manure applications. Adjustments can be made on the basis of a measurement of the topsoil organic matter content, or data from a laboratory anaerobic incubation or from agronomic factors using a computer model. As a guide where measurement is not done, a soil that has a topsoil organic matter content of 10% may release 60-90 kg/ha more potentially available nitrogen than an equivalent soil with 3% organic matter content. However, some soils with an organic matter content of around 10% may release little nitrogen and local knowledge must be used in estimating mineralisable nitrogen.

Add any adjustment for net mineralisable nitrogen to the total of SMN and nitrogen in the crop to give SNS.

Section 3: Using the recommendation tables

Step 4. Identify the SNS Index

SNS = SMN + N in crop + net mineralisable N (kg N/ha)	SNS Index
Less than 60	0
61 – 80	1
81 – 100	2
101 – 120	3
121 – 160	4
161 – 240	5
More than 240	6

Example 1

Spring barley (feed) is to be grown on a light sand soil following sugar beet. Annual rainfall is 650 mm. There have been no organic manures applied or grass grown in the last 5 years.

Select Table B (SNS Indices for moderate rainfall areas). On a light sand soil following sugar beet, the SNS Index is 0. Refer to the spring barley recommendation table on page 111 which gives a recommendation of 110 kg N/ha.

Example 2

Sugar beet is grown on a medium soil after winter wheat. 30 m³/ha of pig slurry (4% DM) was applied in February and incorporated into the soil within 6 hours. Although the average annual rainfall is 650 mm, in an unusually dry winter the excess winter rainfall was found to be 100 mm.

Since the winter was dry, select Table A (SNS Indices for low rainfall areas). On a medium soil after winter wheat, the SNS Index is 1. Refer to the sugar beet recommendation table on page 125 which gives a recommendation of 120 kg N/ha.

Since the pig slurry was applied after harvest of the last crop, its nitrogen contribution must be calculated separately. This manure application provides 86 kg/ha of available nitrogen that is equivalent to inorganic nitrogen fertiliser (see Section 2).

120 – 86 = 34 kg N/ha as fertiliser should be applied.

Section 3: Using the recommendation tables

Example 3

Winter wheat is grown on a medium textured, low organic matter soil after potatoes which received some FYM. Annual rainfall is 750 mm. The soil is sampled in early February and analysed for SMN.

The analysis report shows that the SMN (0-90cm) is 115 kg N/ha and the crop nitrogen content is estimated to be 25 kg N/ha (see Appendix 2). Because the soil contains little organic matter, no extra allowance is made for net mineralisable nitrogen. The SNS is therefore 140 kg N/ha. Refer to any of the SNS Index tables which show that the SNS Index is 4. Refer to the winter wheat recommendation table on page 105, which gives a recommendation of 120 kg N/ha for a medium soil.

Example 4

Winter barley is to be sown following a 3 year pure grass ley which has been managed in the last 2 years using 280 kg/ha/year total nitrogen. An average application of slurry has been applied in early spring each year before taking one cut of silage followed by grazing. The soil is a medium soil in a moderate rainfall area.

The previous grass management is classed as 'High N'. Using Table D2 for medium soils (see page 94), select the category '3-5 year leys, High N, grazed'. The SNS Index appropriate for the winter barley crop is Index 3. The SNS Indices for the next two crops following the winter barley are Index 3 and Index 2 respectively.

Example 5

Winter wheat is to be sown following spring barley that followed a 2 year grazed ley which has been managed using 300 kg/ha/year total nitrogen. The soil is a deep clay in a high rainfall area.

Using Table C, the SNS Index would be 1. Using Table D4, the previous grass management is classed as 'High N' and grazed. The SNS Index from this Table is 2. The higher of these two Indices from Tables C and D4 is 2 and this should be used for the recommendation tables.

Finding the phosphate, potash and magnesium recommendations

Phosphate, potash and magnesium recommendations are based on achieving and maintaining target soil indices for each nutrient in the soil throughout the crop rotation. Soil analysis should be done every 3-5 years. The use of soil analysis as a basis for making fertiliser decisions is described on page 29, and the procedure for taking soil samples in Appendix 3.

The phosphate and potash recommendations shown at Index 2 and 2- respectively are those required to replace the offtake in the yield shown (apart from potatoes where the phosphate recommendation at Index 2 is greater than offtake). The recommendation should be increased or decreased where yields are substantially more or less than this. The amount to apply can be calculated using the expected yield and values for the offtake of phosphate and potash per tonne of yield given in Appendix 5. The larger recommended applications for soils at Index 0 and 1 will also bring the soil to Index 2 over a number of years.

Section 3: Using the recommendation tables

Recommendations are appropriate where the phosphate or potash balance for preceding crops have been close to neutral. Adjustments can be made where the balance for the preceding crop was significantly positive or negative. For example, potatoes can leave a positive phosphate balance so that less than the normally recommended amount might be needed by the following crop. On the other hand, a phosphate or potash 'holiday' can result in a need for greater than normally recommended amounts for following crops.

Recommendations are given as phosphate (P_2O_5), potash (K_2O) and magnesium oxide (MgO). Conversion tables (metric-imperial, oxide-element) are given in Appendix 8.

> ## Example 1
>
> *Soil analysis shows P Index 2 and K Index 1. The next crop to be grown is spring barley, the expected grain yield is 6t/ha and the straw will be baled and removed from the field.*
>
> The table on page 114 recommends 50 kg P_2O_5/ha and 100 kg K_2O/ha.

Other important points to consider when using the recommendation tables are:

- All recommendations are given for the mid-point of each Index. For some crops, there are different recommendations depending on whether the soil is in the lower half (2-) or upper half (2+) of K Index 2.

- Where a soil analysis *value* (as given by the laboratory) is close to the range of an adjacent Index, the recommendation may be reduced or increased slightly taking account of the recommendation given for the adjacent Index. Small adjustments of less than 10 kg/ha are generally not justified.

> ## Example 2
>
> *Soil K analysis is 65 mg/litre which is at the low end of K Index 1 (range 61 and 120 mg/litre K – see Appendix 4). Winter wheat is to be grown, expected grain yield is 8t/ha and straw will be removed.*
>
> The table on page 114 shows a recommendation of 115 kg K_2O/ha for winter wheat (straw removed) at soil K Index 1. This recommendation is for a soil K analysis value of 90 mg/litre, the mid-point of K Index 1. Because the soil is at the bottom of Index 1, it would be more appropriate to apply 130 kg K_2O/ha, a value between that for K Index 0 and 1.

- Where more or less phosphate and potash are applied than suggested in the tables adjustments can be made later in the rotation.

Section 3: Using the recommendation tables

> **Example 3**
>
> *Soil analysis shows P Index 2 and K Index 2, and main crop potatoes are to be grown and expected yield is 65 t/ha tubers.*
>
> The Table on page 123 recommends 170 kg P_2O_5/ha and 300 kg K_2O/ha for a crop yielding 50 t/ha. Using Appendix 5, a crop yielding 65 t/ha will remove:
>
> Phosphate   65 x 1.0 = 65 kg P_2O_5/ha
> Potash   65 x 5.8 = 377 kg K_2O/ha
>
> For phosphate, the recommendation is much larger than the offtake because potatoes are likely to respond to extra phosphate at P Index 2. The surplus phosphate, 105 kg P_2O_5/ha (170 – 65), should be allowed for when deciding on the phosphate application for the next crop(s) grown in the rotation.
>
> For potash, the recommendation is less than the offtake and the deficit (377 – 300) should be made good later in the crop rotation.

- Where organic manure is applied it is important to calculate the quantity of each plant available nutrient added in the manure (see Section 2) and adjust the amount of fertiliser by this amount. Where organic manures are applied frequently it is essential to calculate the immediate and residual plant-available amount of each nutrient and adjust the fertiliser applied appropriately. Allowing for the nutrients in manure reduces the need for fertiliser, improves farm profits and reduces the risk of nutrient pollution of water.

- Construct a nutrient balance sheet for each field and ensure that the phosphate and potash offtake is balanced by an equivalent application of phosphate and potash on Index 2 soils and check that the soil is maintained at Index 2 by soil analysis every 3-5 years.

Finding the sulphur and sodium recommendations

Sulphur and sodium recommendations are given for each crop where appropriate because they are not required by all crops and in all parts of England and Wales. Farmers are advised to monitor the sulphur requirements of their crops because the risk of sulphur deficiency is increasing as sulphur inputs to agricultural soils decline. Organic manures can supply useful amounts of sulphur (see Section 2).

All sulphur recommendations are given as SO_3 and sodium recommendations as Na_2O. Conversion tables (metric-imperial, oxide-element) are given in Appendix 8.

Section 3: Using the recommendation tables

Selecting the most appropriate fertiliser

For a single nutrient, the recommended amount can be applied using a straight fertiliser. Where more than one nutrient is required a compound or blended fertiliser can be used. In this case, the compound or blend selected will depend on the ratio of the nutrients in the fertiliser and the amount applied should give as near the recommended amount of each nutrient as possible. Often it will not be possible to exactly match the recommendations with available fertilisers. In most cases, the first priority is to get the amount of nitrogen correct because crops respond most to nitrogen. Slight variation in the rates of phosphate or potash will have less effect on yield, especially on Index 2 soils, and any discrepancy can be corrected in fertiliser applications to future crops. More information on fertiliser selection is given on page 46.

Calculating the amount of fertiliser to apply

The content of each nutrient in a fertiliser is given as a percentage of N, P_2O_5 and K_2O.

> **Example**
>
> Winter wheat grown on a P Index 2, K Index 1 soil and expected to yield 8 t/ha grain requires 200 kg N/ha, 65 kg P_2O_5/ha and 114 kg K_2O/ha. A 20: 10: 10 NPK compound fertiliser is available and 100 kg contains 20 kg N, 10 kg P_2O_5 and 10 kg K_2O. Applied at 1000 kg/ha this fertiliser will supply 200 kg N/ha, 100 kg P_2O_5/ha and 100 kg K_2O/ha.
>
> The surplus 35 kg P_2O_5/ha can be allowed for later in the rotation.
>
> The deficit of 15 kg K_2O/ha can be ignored.

Metric to imperial conversion tables are given in Appendix 8. If applying liquid fertilisers, manufacturers can supply tables which convert kg/ha of nutrient to litres/ha of product.

Section 4: Arable and forage crops

	Page
Checklist for decision making	104
Wheat, autumn and early winter sown – nitrogen	105
Barley, winter sown – nitrogen	107
Oats, rye and triticale, winter sown – nitrogen	109
Wheat, spring sown – nitrogen	110
Barley, spring sown – nitrogen	111
Oats, rye and triticale, spring sown – nitrogen	112
All cereals – phosphate, potash, magnesium and sulphur	113
Oilseed rape, autumn and winter sown – nitrogen	115
Oilseed rape and linseed, spring sown – nitrogen	117
Oilseed rape and linseed – phosphate and potash	118
Peas (dried and vining) and beans	120
Potatoes – nitrogen	121
Potatoes – phosphate, potash and magnesium	123
Sugar beet	125
Forage maize	127
Other forage crops	129
Ryegrass grown for seed	131

For each crop, recommendations for nitrogen (N), phosphate (P_2O_5) and potash (K_2O) are given in kilograms per hectare (kg/ha). Magnesium (MgO), sulphur (as SO_3) and sodium (as Na_2O) recommendations, also in kg/ha, are given where these nutrients are needed. Recommendations are given for the rate and timing of fertiliser application.

The recommendations are based on the nutrient requirements of the crop being grown, making allowance for the nutrients already contained in the soil. Section 3 gives instructions on how to use the tables, including how to determine the Soil Nitrogen Supply (SNS) Index.

More background on the principles underlying the recommendations is given in Section 1.

Section 4: Arable and forage crops

Checklist for Decision Making

Individual decisions for fertiliser use must be made for each field on the farm. Where more than one crop is grown in a field, these areas must be considered individually.

1. Confirm the crop to be grown and the intended market. Identify any crop quality requirements required by this market. For the purposes of this guide, winter sown is defined as sown before 1st January and spring sown as in January or after.

2. Identify the dominant soil type in the cropped area (see page 86 and Appendix 1).

3. Carry out soil analysis for pH, P, K and Mg every 3-5 years (see page 35). Target values to maintain in arable rotations are:

    Soil pH 6.5 (5.8 on peat soils)
    Soil P Index 2
    Soil K lower Index 2 (2-)
    Soil Mg Index 2

4. Identify the Soil Nitrogen Supply (SNS) Index of the field either by carrying out analysis for soil mineral nitrogen (the Measurement Method) (see page 95) or by using the Field Assessment Method based on previous cropping, previous fertiliser and manure use, soil type and winter rainfall (see page 86). Sampling and analysis for SMN is recommended where nitrogen residues are expected to be moderate or high (e.g. following previous manure use or crops receiving high rates of nitrogen fertiliser or following a dry winter).

5. Calculate the nutrients that will be available for crop uptake from organic manures that have been applied since harvest of the previous crop, or which will be applied to the crop being grown (see Section 2). Deduct these nutrients from the recommended rates given in the tables.

6. Decide on the strategy for phosphate and potash use. This will be building up, maintaining or running down the soil Index (see pages 38-41). Allow for any surplus or deficit of phosphate or potash applied to previous crops in the rotation.

7. Calculate the amount of phosphate and potash removed in the harvested crop according to expected crop yield (see pages 38-41 and Appendix 5). This is the amount of these nutrients that must be replaced in order to maintain the soil at the current Index. Remember that some of these nutrients will also be removed in straw.

8. Using the tables, decide on the required rate of each nutrient. Decide on the optimum timings for fertiliser application, then find the best match for these applications using available fertilisers (see page 102).

9. Check that the fertiliser spreader or sprayer is in good working order and has been recently calibrated (see page 49).

10. Keep an accurate record of the fertilisers and organic manures applied.

Section 4: Arable and forage crops

Wheat, Autumn and Early Winter sown – Nitrogen

| | SNS Index | | | | | | |
	0	1	2	3	4	5	6
	kg N/ha						
Light sand soils	160	130	100	70	40	0-40	0-40
Shallow soils	280	240	210	180	140	80	0-40
Medium soils	250	220	190	160	120	60	0-40
Deep Clay soils	250	220	190	160	120	60	0-40
Deep Silty soils	220	190	160	130	100	40	0-40
Organic soils				120	80	40-80	0-40
Peat						0-60	

Nitrogen

There is no requirement for seedbed nitrogen.

Depending on the total nitrogen requirement and crop development, it will often be appropriate to apply nitrogen at the following timings.

- Less than 120 kg N/ha:
  Apply all the recommended amount as a single dressing by early stem extension but not before early April.
- Over 120 kg N/ha:
  Apply about 40 kg N/ha between mid-February and mid-March except where
  - There is a low risk of take-all and
  - Shoot numbers are very high. Well-tillered crops do not need nitrogen at this stage. Crops with too many tillers will be prone to lodging and higher disease levels.

The balance of the application should be applied in one or two dressings during early stem extension. Where more than 120 kg N/ha remains to be applied, half should be applied at the start of stem extension (not before April), and half at least 2 weeks later (not after early May).

Yield variation and use of grain nitrogen concentration

Research has shown that the main causes of yield variation between fields (soil type, rotational position, sowing date or variety) are not associated with variations in the economic rate of nitrogen fertiliser.

Farm nitrogen strategies for wheat should be assessed periodically using information on grain nitrogen concentration. This is a better guide than yield to indicate whether nitrogen fertiliser use is correct. Where there is a consistent trend of low grain %N over several years then nitrogen rates should be increased. On the other hand, when yields are consistently small, it is difficult to justify the amounts of nitrogen shown in the table without good supporting evidence (e.g. from grain %N). Grain %N at the economic optimum rate of nitrogen is about 1.9% N (100% dry matter basis) for feed wheat and 2.1% nitrogen for breadmaking wheat. Where concentrations

Section 4: Arable and forage crops

in yields from a number of adjacent fields are consistently above or below these values during several years, nitrogen fertiliser application rates should be adjusted down or up by 30 kg N/ha per 0.1% difference in grain %N.

To convert grain %protein to grain %N, divide by 5.7. Both N and protein are both reported on a 100% dry matter basis.

The effect of economic changes on nitrogen rates

The recommendations in the tables for wheat and barley are based on a breakeven ratio of 5.0 (cost of fertiliser nitrogen as £/kg N divided by value of grain as £/kg) (see page 32). If the price of nitrogen or the price of grain changes, use the table below to decide on an amount to add to or subtract from the fertiliser nitrogen application.

	Fertiliser N content (%)	Fertiliser Cost (£/tonne PRODUCT)					
Ammonium Nitrate	34.5%	£138	£207	£276	£345	£414	£483
Urea[a]	46.0%	£184	£276	£368	£460	£552	£644
Urea-Ammonium Nitrate liquid[a]	28.0%	£112	£168	£224	£280	£336	£392
Cost of Fertiliser Nitrogen	£/kg N	£0.40	£0.60	£0.80	£1.00	£1.20	£1.40
Change to recommended N for all Cereals (kg N/ha)							
Grain sale price (£/tonne)	50	-30	-60	-80	-100	-110	-130
	75	0	-30	-50	-70	-80	-90
	100	10	-10	-30	-40	-60	-70
	125	20	0	-10	-30	-40	-50
	150	30	10	0	-20	-30	-40
	175	40	20	0	-10	-20	-30
	200	40	30	10	0	-10	-20
	225	40	30	20	10	0	-10
	250	50	30	20	10	0	-10
	275	50	40	30	20	10	0

a. See notes on efficiency of use of different kinds of applied nitrogen page 30

Crops grown for breadmaking

In some circumstances, an application of nitrogen in addition to that recommended above may be economically worthwhile to boost the grain protein concentration. Typically an application of an extra 60 kg N/ha will increase grain protein by 1.1%. Application of soil-applied additional nitrogen during stem extension may give a small yield increase as well as an increase in grain protein. Application as a foliar urea spray during, but no later than, milky ripe stage will result in a larger increase in grain protein content but cannot be expected to increase yield.

Don't forget to deduct nutrients applied as organic manures (see Section 2)

Section 4: Arable and forage crops

Barley, winter sown – Nitrogen

	SNS Index						
	0	1	2	3	4	5	6
	kg N/ha						
Feed barley							
Light sand soils	150	120	90	60	30-60	0-30	0
Shallow soils	210	190	150	120	60	20-60	0-20
Medium and deep clay soils	190	170	140	110	60	20-60	0-20
Deep fertile silty soils	160	140	110	70	30	0-20	0
Organic soils				110	60	0-40	0
Peaty soils						0-40	
Malting barley (up to 1.8% grain N)							
Light sand soils	120	80	40	0-40	0	0	0
Other mineral soils	160	130	100	70	0-40	0	0
Organic soils				70	0-40	0	0
Peaty soils						0	

Nitrogen – feed barley

There is no requirement for seedbed nitrogen.

Where the total nitrogen rate is less than 100 kg N/ha, apply this amount as a single dressing by early stem extension, but not before late March.

Where the total nitrogen rate is 100 kg N/ha or more, split the dressing with 40 kg N/ha in mid-February/early March and the rest by early stem extension, but not before late March.

These recommendations assume appropriate measures are taken to control lodging (e.g. choice of variety, use of plant growth regulator). Reduce the recommendation by 25 kg N/ha if the lodging risk is high.

Nitrogen – malting barley

Careful judgement of the nitrogen rate is important to ensure that the grain nitrogen concentration is neither too high nor too low for the requirement of the target market. Previous experience and consultation will be important in deciding the nitrogen rate to use. Where quality premiums are expected to be low, applying a slightly higher nitrogen rate will maximise the yield potential of the crop.

Section 4: Arable and forage crops

On light sand soils, the recommended rate assumes that the grain yield will be no more than 6.0-6.5 t/ha. Low yields are normal on these soils due to summer drought stress. Where there is a realistic expectation of a yield greater than this, the nitrogen recommendation for light sand soils may be increased by 10-20 kg N/ha.

Where the total nitrogen rate is 100 kg N/ha or more, split the dressing with 40 kg N/ha in mid February-early March and the rest by end March. Where the nitrogen rate is below 100 kg N/ha, apply a single dressing by end-March.

Where the target grain %N is below 1.8%, the nitrogen rate may need to be reduced by up to 30 kg N/ha taking account of previous farm experience. This nitrogen should all be applied by mid March.

The effect of economic changes on nitrogen rates

The recommendations in the table are based on a breakeven ratio of 5.0 (see page 32). If the price of nitrogen or the price of grain changes, use the winter sown wheat table above to decide on an amount to add to or subtract from the nitrogen application.

Don't forget to deduct nutrients applied as organic manures (see Section 2)

Section 4: Arable and forage crops

Oats, Rye and Triticale, winter sown – Nitrogen

	SNS Index						
	0	1	2	3	4	5	6
	kg N/ha						
Light sand soils	110	70	20-50	0-20	0	0	0
All other mineral soils	150	120	90	60	30	0-20	0
Organic soils				60	30	0-20	0
Peaty soils						0-20	

Timing of application

Where the total nitrogen rate is less than 100 kg N/ha, apply as a single dressing by early stem extension, but not before late March. Where the total nitrogen rate is 100 kg N/ha or more, split the dressing with 40 kg N/ha in mid February-early March and the rest by early stem extension, but not before late March.

These recommendations assume appropriate measures are taken to control lodging (e.g. choice of variety, use of plant growth regulator). Reduce the recommended rate by 25 kg N/ha if lodging risk is high.

The effect of economic changes on nitrogen rates

The recommendations in the table are based on a breakeven ratio of 5.0 (see page 32). If the price of nitrogen or the price of grain changes, use the winter sown wheat table above to decide on an amount to add to or subtract from the nitrogen application.

Don't forget to deduct nutrients applied as organic manures (see Section 2)

Section 4: Arable and forage crops

Wheat, spring sown – Nitrogen

	SNS Index						
	0	1	2	3	4	5	6
	kg N/ha						
Light sand soils	160	130	100	70	40	0-40	0
All other mineral soils	210	180	150	120	70	40	0-40
Organic soils				120	70	40	0-40
Peaty soils						0-40	

Nitrogen

For crops drilled before March, apply nitrogen at early stem extension but not before early April or after early May. For amounts greater than 70 kg N/ha, apply 40 kg N/ha of the total in the seedbed except on light sand soils. On these soils apply 40 kg N/ha at the 3 leaf stage but not before March.

For late-drilled crops, all the nitrogen can be applied in the seedbed except on light sand soils where amounts greater than 70 kg N/ha should be split with 40 kg N/ha in the seedbed and the remainder by the 3 leaf stage.

Bread making varieties

In some circumstances, an application of nitrogen in addition to that given in the table may be economically worthwhile to boost the grain protein concentration. Typically an application of an extra 40 kg N/ha will increase grain protein by 1.1%. Application of this additional nitrogen during stem extension may give a small yield increase as well as an increase in grain protein. Application as a foliar urea spray during, but not later than, milky ripe stage will result in a larger increase in grain protein content but cannot be expected to increase yield.

The effect of economic changes on nitrogen rates

The recommendations in the table are based on a breakeven ratio of 5.0 (see page 32). If the price of nitrogen or the price of grain changes, use the winter sown wheat table above to decide on an amount to add to or subtract from the nitrogen application

Don't forget to deduct nutrients applied as organic manures (see Section 2)

Section 4: Arable and forage crops

Barley, spring sown – Nitrogen

	SNS Index						
	0	1	2	3	4	5	6
	kg N/ha						
Feed							
Light sand soils	110	80	50	30	0-30	0	0
Other mineral soils	160	140	110	70	30	0-30	0
Organic soils				70	30	0-30	0
Peaty soils						0-30	
Malting barley (up to 1.8% grain N)							
Light sand soils	100	70	40	0-40	0	0	0
Other mineral soils	140	120	80	50	0-40	0	0
Organic soils				50	0-40	0	0
Peaty soils						0	

Nitrogen – feed barley

For crops drilled before March, apply nitrogen at early stem extension but not before early April or after early May. For amounts greater than 70 kg N/ha, apply 40 kg N/ha of the total in the seedbed except on light sand soils. On these soils apply 40 kg N/ha at the 3 leaf stage but not before March.

For late-drilled crops, all the nitrogen can be applied in the seedbed except on light sand soils where amounts greater than 70 kg N/ha should be split with 40 kg N/ha in the seedbed and the remainder by the 3 leaf stage.

Nitrogen – malting barley

Careful judgement of the nitrogen rate is important to ensure that the grain nitrogen concentration is neither too high nor too low for the requirement of the target market. Previous experience and consultation will be important in deciding the nitrogen rate to use. Where quality premiums are expected to be low, use of a slightly higher nitrogen rate will maximise the yield potential of the crop.

Apply all the nitrogen by early stem extension but not after end March. Where the target grain %N is below 1.8%, the nitrogen rate may need to be reduced by up to 20-30 kg N/ha taking account of previous farm experience. This nitrogen should all be applied by mid March.

The effect of economic changes on nitrogen rates

The recommendations in the table are based on a breakeven ratio of 5.0 (see page 32). If the price of nitrogen or the price of grain changes, use the winter sown wheat table above to decide on an amount to add to or subtract from the nitrogen application

Don't forget to deduct nutrients applied as organic manures (see section 2)

Section 4: Arable and forage crops

Oats, Rye and Triticale, spring sown – Nitrogen

	SNS Index						
	0	1	2	3	4	5	6
	kg N/ha						
Light sand soils	90	60	30	0-30	0	0	0
Other mineral soils	140	110	70	40	0-30	0	0
Organic soils				40	0-30	0	0
Peaty soils						0	

The effect of economic changes on nitrogen rates

The recommendations in the table are based on a breakeven ratio of 5.0 (see page 32). If the price of nitrogen or the price of grain changes, use the winter sown wheat table above to decide on an amount to add to or subtract from the nitrogen application.

Don't forget to deduct nutrients applied as organic manures (see Section 2)

Section 4: Arable and forage crops

All Cereals – Phosphate, Potash, Magnesium and Sulphur

The amounts of phosphate and potash are appropriate to the grain yields shown.
The amounts of phosphate and potash shown at the target Index, P Index 2 and K Index 2-, are needed to replace the offtake in the yield of grain shown and maintain the soil at the target Index. The upper half of the Table shows the maintenance applications for the yield shown for each cereal when straw is incorporated. The lower half of the Table shows the maintenance applications when straw is removed but not weighed, the extra amount of phosphate and potash is based on the grain yield. Where the weight of straw removed is known, use the amounts of phosphate and potash per tonne straw shown in Appendix 5 to calculate the amounts removed and apply these amounts as the maintenance dressing. The phosphate and potash recommendations at target or lower Indices can be adjusted if yields are likely to be larger or smaller than those shown in the table, by multiplying the difference in expected grain yield by the phosphate and potash content per tonne of grain yield using the appropriate value for where straw is either incorporated or removed as given in Appendix 5. For example, at P Index 1, the phosphate recommendation for wheat with an expected yield of 10 t/ha where straw is incorporated is $90 + (2 \times 7.8) = 106$ kg/ha. No adjustment for yield should be made where the soil Index is higher than target.

Crops grown on soil at Index 0 and 1 would be expected to respond to the extra amounts of phosphate and potash shown in the table below. Also over a number of years, these extra amounts of fertiliser will help to raise most soils, except light sands, to Index 2. At Index 3 and above no phosphate and potash need be applied for a few years but check soil analyses regularly.

At Index 2, phosphate and potash can be applied when convenient during the year but at Index 0 and 1, they should be applied and worked into the seedbed. To avoid damage to germinating seedlings do not combine drill more than 150 kg/ha of nitrogen plus potash on sandy soils.

Section 4: Arable and forage crops

	P or K Index			
	0	1	2	3 and higher
	kg/ha			
Straw ploughed in/incorporated				
Winter wheat, winter barley (8 t/ha)				
Phosphate (P$_2$O$_5$)	120	90	60	0
Potash (K$_2$O)	105	75	45 (2-) 20 (2+)	0
Spring wheat, spring barley, rye, triticale (6 t/ha)				
Phosphate (P$_2$O$_5$)	105	75	45	0
Potash (K$_2$O)	95	65	35 (2-) 0 (2+)	0
Winter and spring oats (6 t/ha)				
Phosphate (P$_2$O$_5$)	105	75	45	0
Potash (K$_2$O)	95	65	35 (2-) 0 (2+)	0
Straw removed				
Winter wheat, winter barley (8 t/ha)				
Phosphate (P$_2$O$_5$)	125	95	65	0
Potash (K$_2$O)	145	115	85 (2-) 55 (2+)	0
Spring wheat, spring barley, rye, triticale (6 t/ha)				
Phosphate (P$_2$O$_5$)	110	80	50	0
Potash (K$_2$O)	130	100	70 (2-) 40 (2+)	0
Winter and spring oats (6 t/ha)				
Phosphate (P$_2$O$_5$)	115	85	55	0
Potash (K$_2$O)	165	135	105 (2-) 75 (2-)	0

Magnesium

At Mg Index 0, magnesium fertiliser should be applied every 3-4 years at 50 to 100 kg MgO/ha (see page 42).

Sulphur

Cereals are becoming more responsive to sulphur as atmospheric deposition of sulphur declines. See page 43 for more details and a map showing current deposition and areas where deficiency could occur. Where deficiency has been recognised or is expected, 25-40 kg SO$_3$/ha as a sulphate-containing fertiliser should be applied in early spring before the start of stem extension.

Sulphur deficiency can be diagnosed by analysing whole plant shoots sampled at stem extension. A value of less than 0.2% S in dry-matter indicates deficiency. At this stage of growth there is little opportunity to correct any deficiency, but identifying deficiency allows remedial action to be taken for subsequent crops.

Don't forget to:

make allowance for nutrients applied in organic manures (see Section 2)

ensure the phosphate and potash offtake is balanced by application on Index 2 soils and check that the soil is maintained at Index 2 by soil sampling every 3-5 years

Section 4: Arable and forage crops

Oilseed rape, Autumn and Winter Sown – Nitrogen

	SNS Index						
	0	1	2	3	4	5	6
	kg N/ha						
Autumn	30	30	30	0	0	0	0
Spring							
All mineral soils	220	190	160	120	80	40-80	0-40
Organic soils				120	80	40-80	0-40
Peaty soils						40-80	40-80

Timing of application

Autumn nitrogen can be applied to the seedbed or as a top-dressing to encourage autumn growth but research suggests that crops sown after early September are unlikely to respond.

Where the nitrogen rate is less than 100 kg N/ha, apply the whole dressing during late February-early March at the start of spring growth.

For amounts of 100 kg N/ha or greater, split the dressing with half at this time and the remainder by late March-early April.

Yield variation

The recommendations are for crops yielding 3.5 t/ha. Where previous experience of growing the crop indicates that yields above 3.5 t/ha can be expected realistically, the recommended rate may be increased by up to 30 kg N/ha per half tonne of yield up to an expected yield of 4.5 t/ha. This adjustment should be used with caution because applying too much early nitrogen to crops with large canopies can increase lodging and possibly reduce yield.

The effect of economic changes on nitrogen rates

The recommendations in the table are based on a breakeven ratio of 2.5 (see page 32). If the price of nitrogen or the value of grain changes, use the table below to decide on an amount to add or subtract from the nitrogen application.

Don't forget to deduct nutrients applied as organic manures (see section 2)

Section 4: Arable and forage crops

	Fertiliser N content (%)	Fertiliser Cost (£/tonne Product)					
Ammonium Nitrate	34.5%	£138	£207	£276	£345	£414	£483
Urea[a]	46.0%	£184	£276	£368	£460	£552	£644
Urea-Ammonium Nitrate Liquid[a]	28.0%	£112	£168	£224	£280	£336	£392
Cost of Fertiliser Nitrogen	£/kg N	£0.40	£0.60	£0.80	£1.00	£1.20	£1.40
Change to recommended N for Oilseed Rape (kg N/ha)							
Grain sale price (£/tonne)	200	40	0	-30	-50	-70	-80
	225	50	10	-20	-40	-60	-70
	250	60	20	-10	-30	-50	-60
	275	70	30	0	-20	-40	-50
	300	80	40	10	-10	-30	-40
	325	80	50	20	-0	-20	-30
	350	90	50	30	0	-10	-30
	375	100	60	30	10	-10	-20
	400	100	70	40	20	0	-10
	425	110	70	40	20	10	-10
	450	110	80	50	30	10	-0

a. See notes on efficiency of use of different kinds of applied nitrogen page 30

Section 4: Arable and forage crops

Oilseed Rape and Linseed, Spring Sown – Nitrogen

	SNS Index						
	0	1	2	3	4	5	6
	kg N/ha						
Spring oilseed rape							
Light sand soils	120	80	50	0-40	0	0	0
Other mineral soils	150	120	80	50	0-40	0	0
Organic soils				50	0-40	0	0
Peaty soils						0-40	
Spring linseed							
Light sand soils	80	50	0-40	0	0	0	0
All other mineral soils	100	80	50	0-40	0	0	0
Organic soils				0-40	0	0	0
Peaty soils						0	

Nitrogen

Apply all the nitrogen in the seedbed. On light sand soils where the total rate is more than 80 kg N/ha, the dressing should be split with 50 kg N/ha in the seedbed and the remainder by early May.

The recommendations in the table are based on a breakeven ratio of 2.5 (see page 33). If the price of nitrogen or the value of grain changes, use the table under winter sown oilseed rape to decide on an amount to add to or subtract from the nitrogen application.

Don't forget to deduct nutrients applied as organic manures (see Section 2)

Section 4: Arable and forage crops

Oilseed rape and Linseed – Phosphate, Potash, Magnesium and Sulphur

The amounts of phosphate and potash are appropriate to the seed yields shown. For each crop in the Table the amounts of phosphate and potash shown at the target Index, P Index 2 and K Index 2-, are required to replace the offtake in the yield shown and maintain the soil at the target Index. The recommendation at target or lower Indices can be adjusted if the yields are likely to be larger or smaller than those shown by multiplying the difference in expected yield by the phosphate and potash content per tonne of seed given in Appendix 5. For example, at P Index 1, the phosphate recommendation for winter oilseed rape with an expected yield of 4.5 t/ha is $80 + (1 \times 14) = 94$ kg P_2O_5/ha. No adjustment for yield should be made where the soil Index is higher than target.

Crops grown on soil at Index 0 and 1 would be expected to respond to the extra amounts of phosphate and potash shown in the table below. Also, over a period of years, this additional amount of fertiliser will help raise the soil to Index 2.

Apply phosphate and potash when convenient during the year except on Index 0 and 1 soils when it should be applied and worked into the seedbed.

	P or K Index			
	0	1	2	3 and higher
	kg/ha			
Winter oilseed rape (3.5 t/ha)				
Phosphate (P_2O_5)	110	80	50	0
Potash (K_2O)	100	70	40 (2-) 20 (2+)	0
Spring oilseed rape (2 t/ha) or Linseed (1.5 t/ha)				
Phosphate (P_2O_5)	90	60	30	0
Potash (K_2O)	80	50	20 (2-) 0 (2+)	0

Magnesium

At Mg Index 0 and 1, magnesium at 50 to 100 kg MgO/ha should be applied every three or four years (see page 42 for details).

Sulphur

Oilseed rape will respond to an application of sulphur on all mineral soils. Spring crops may be less susceptible to sulphur deficiency than winter crops. See page 43 for more details and a map showing current deposition and areas where deficiency could occur. Where deficiency has been recognised or is suspected, 50-75 kg SO_3/ha as a sulphate containing fertiliser should be applied in early spring.

Section 4: Arable and forage crops

Sulphur deficiency can be diagnosed by analysing young fully expanded leaves at early flowering stage. Critical values of less than 0.4% S in dry-matter or an N: S ratio of more than 17: 1 indicate deficiency. At this stage of growth there is little opportunity to correct any deficiency, but identifying deficiency allows remedial action to be taken for subsequent crops. The leaf sulphate: malate ratio test can predict potential deficiency at an earlier stage of growth. Sulphur deficiency symptoms include stunting, interveinal yellowing of middle and upper leaves and pale flower petals.

Don't forget to:

make allowance for applied in organic manures (see Section 2)

ensure the phosphate and potash offtake is balanced by application on Index 2 soils and check that the soil is maintained at Index 2 by soil sampling every 3 – 5 years

Section 4: Arable and forage crops

Peas (dried and vining) and Beans

The amounts of phosphate and potash are appropriate to pea yields of 4 t/ha and bean yields of 3.5 t/ha. Where yields are likely to be greater or smaller, phosphate and potash applications should be adjusted accordingly. Appendix 5 gives typical values of the phosphate and potash content in crop material per tonne of yield.

	SNS, P, K or Mg Index						
	0	1	2	3	4	5	6
	kg N/ha						
Nitrogen (N)	0	0	0	0	0	0	0
Phosphate (P_2O_5)	100	70	40	0	0	0	0
Potash (K_2O)	100	70	40 (2-) 20 (2+)	0	0	0	0
Magnesium (MgO)	100	50	0	0	0	0	0

Phosphate and potash

Seedbed phosphate and potash is only needed at Index 0 and 1.

Sulphur

Peas may suffer from sulphur deficiency on sensitive soil types (see page 43). Where deficiency is possible, apply 25 kg SO_3/ha.

Don't forget to:

make allowance for nutrients applied in organic manures (see Section 2)

ensure the phosphate and potash offtake is balanced by application on Index 2 soils and check that the soil is maintained at Index 2 by soil sampling every 3 – 5 years

Section 4: Arable and forage crops

Potatoes – Nitrogen

The recommendations below provide general guidance only. Because of the very wide range of varietal characteristics and quality requirements for different market outlets, guidance from a FACTS Qualified Adviser should be used when making decisions for specific crops.

The SNS Index below should be used together with the anticipated length of growing season and variety group to determine the appropriate range of nitrogen rates.

Length of growing season[a] and variety group[b]		SNS Index		
		0 and 1	2, 3 and 4	5 and 6
		kg N/ha		
<60 days	– Variety group 1	100-140	70-110	40-60
	– Variety group 2	80-120	50-80	0-40
	– Variety group 3	60-100	40-70	0-40
	– Variety group 4	N/A	N/A	N/A
60-90 days	– Variety group 1	160-210	130-160	90-120
	– Variety group 2	100-160	60-120	40-80
	– Variety group 3	60-140	40-100	0-60
	– Variety group 4	40-80	20-40	0-40
90-120 days	– Variety group 1	220-270	190-220	150-180
	– Variety group 2	150-220	110-160	80-120
	– Variety group 3	110-180	80-100	40-60
	– Variety group 4	80-140	40-60	0-40
>120 days	– Variety group 1	N/A	N/A	N/A
	– Variety group 2	190-250	150-180	120-150
	– Variety group 3	150-210	120-140	80-100
	– Variety group 4	100-180	60-80	20-40

N/A= Not Applicable
a. 50% emergence to haulm death
b. Examples of varieties to in each variety group are as follows:

Group 1	Short haulm longevity (Determinate varieties)	Accord, Estima, Maris Bard, Rocket and Premiere.
Group 2	Medium haulm longevity (Partially determinate varieties)	Atlantic, Lady Rosetta, Marfona, Maris Peer, Nadine, Saxon, Shepody and Wilja.
Group 3	Long haulm longevity (Indeterminate varieties)	Maincrop varieties such as Desiree, Fianna, Hermes, King Edward, Maris Piper, Rooster, Russet Burbank, Pentland Dell, Pentland Squire and Saturna.
Group 4	Very long haulm longevity	Cara and Markies.

June 2010

Section 4: Arable and forage crops

Nitrogen increases yield by prolonging haulm life. It has no effect on tuber numbers and consequently, where it gives an increase in yield, the mean tuber size will be greater.

Factors that may influence the nitrogen rate

The recommendations assume that loss of ground cover should begin close to the time of defoliation and harvest. If crops are planted later than intended and the defoliation date remains unaltered, this will reduce the length of the growing season which will justify a reduction in the nitrogen application rate.

Excess applications of nitrogen can:

- Increase haulm size, delay natural senescence and create difficulties with desiccation.
- Delay achievement of skin set.
- Affect achievement of target dry matter levels.

Timing of nitrogen application

If top dressing is planned for management reasons or to reduce the risk of leaching for crops grown on light sand and shallow soils, apply about two-thirds of the nitrogen recommendation in the seedbed and the remainder shortly after emergence.

For other crops, apply all of the nitrogen recommendation in the seedbed.

The effect of irrigation

There is no difference in the nitrogen recommendation between irrigated crops and those which are not irrigated.

Placement of nitrogen in bed systems

The same recommendations should be used for bed as well as ridge and furrow systems, and where nitrogen fertiliser is placed.

Don't forget to deduct nutrients applied as organic manures (see Section 2)

Section 4: Arable and forage crops

Potatoes – Phosphate, Potash and Magnesium

The amounts of phosphate and potash shown at Index 2 are those recommended to achieve a **total yield of 50 t/ha.** The phosphate recommendations are intended to achieve optimum yield and should not be adjusted even if larger or smaller yields than 50 t/ha are expected. The potash recommendation at target or lower Indices can be adjusted when yield is likely to be larger or smaller than 50 t/ha by multiplying the difference in expected yield by the potash content per tonne yield given in Appendix 5. For example, at K Index 1, the potash recommendation for an expected yield of 70 t/ha is 330 + (20 x 5.8) = 446 kg K_2O/ha. No adjustment for yield should be made where the soil Index is higher than target.

Crops grown on soil at Index 0 and 1 would be expected to respond to the extra amounts of phosphate, potash and magnesium shown in the table below.

	P, K or Mg Index				
	0	1	2	3	4 and higher
	kg N/ha				
Phosphate (P_2O_5)	250	210	170	100	0
Potash (K_2O)	360	330	300	150	0
Magnesium (MgO)	120	80	40	0	0

The amount of phosphate recommended for soils at P Index 2 or 3 is more than sufficient to replace the phosphate removed by a 50 t/ha crop (about 50 kg P_2O_5). **The surplus phosphate will help to maintain the soil at a target P Index 2 for an arable crop rotation and should be allowed for when assessing the need for phosphate of one or more following crops.** On soils at P Index 0 and 1 the surplus phosphate will help increase the soil P Index and no allowance should be made when deciding the phosphate requirement of a subsequent crop. On soils at P Index 2 or below a large proportion of the phosphate should be water-soluble.

The amount of potash recommended at K Index 2 will only replace the amount removed by a 50 t/ha crop and potash should be applied for the next crop in the rotation to maintain the soil at K Index 2. The extra amounts of potash shown for K Index 0 and 1 soils will slowly increase the soil K Index.

Timing of application

All the phosphate should be applied in the spring and either worked into the seedbed or placed at planting.

Where more than 300 kg K_2O/ha is required, apply half in late autumn/winter and half in spring. On light sandy soils, all the potash fertiliser should be applied after ploughing and no sooner than late winter. Large amounts of potash can sometimes reduce tuber dry matter content. Where this occurs the decrease may be smaller when muriate of potash (MOP, potassium chloride) is replaced by sulphate of potash (SOP, potassium sulphate).

Section 4: Arable and forage crops

These recommendations should be used for both bed and ridge and furrow systems. Where fertiliser is placed, a small reduction in the recommended rate of phosphate and potash could be considered.

Don't forget to:

make an allowance for nutrients applied in organic manures (see Section 2)

ensure the potash offtake is balanced by application of potash fertiliser on Index 2 soils, and check that the soil is maintained at Index 2 for both phosphate and potash by soil sampling every 3 – 5 years

Section 4: Arable and forage crops

Sugar Beet

Nitrogen

The recommendations do not vary with yield. Nitrogen fertilisers should be applied in spring, 30 – 40 kg/ha of the total N required immediately after drilling and the remainder when all the beet seedlings have emerged.

	SNS Index				
	0 and 1	2	3	4	5
	kg N/ha				
All mineral soils	120	100	80	0	0
Organic soils				40	0
Peaty soils					0

If in doubt about the appropriate SNS Index, seek advice from a FACTS Qualified Adviser.

Phosphate, potash, magnesium and sodium

The amounts of phosphate and potash shown at target Index 2 are needed to replace the offtakes in a **60 t/ha crop** (with tops ploughed in) and maintain the soil at the target Index. The phosphate and potash recommendations at target or lower Indices can be adjusted if yields are likely to be larger or smaller than 60 t/ha by multiplying the difference in expected yield by the phosphate and potash content per tonne of yield given in Appendix 5. For example, at P Index 1, the recommendation for an expected yield of 70 t/ha where tops are incorporated is 80 + (10 x 0.8) = 88 kg P_2O_5/ha. Alternatively for potash, growers can access their factory-determined estimates of the amounts of potash removed in their delivered crops from British Sugar Online as a guide to application rates on Index 2 soils.

Crops grown on soil at Index 0 and 1 would be expected to respond to the extra amounts of phosphate, potash and magnesium shown in the table below. Also over a number of years, these extra amounts of fertiliser will help to raise most soils, except light sands, to Index 2.

Section 4: Arable and forage crops

	P, K or Mg Index				
	0	1	2	3	4 and higher
	kg N/ha				
Phosphate (P$_2$O$_5$)	110	80	50	0	0
Potash (K$_2$O)	160	130	100	0	0
Magnesium (MgO)	150	75	0	0	0
Na$_2$O (use K Index)[a]	200	200	100	0	0

a. Sodium can partly replace potash in the nutrition of sugar beet when soils contain too little crop-available potash. An application of 200 kg Na$_2$O/ha is recommended for beet grown on soils at K Index 0 and 1. On K Index 2 soils it is only necessary to apply 100 kg Na$_2$O/ha when the soil contains less than 25 mg Na/kg. Fen peats, silts and clays usually contain sufficient sodium and no fertiliser sodium is recommended. Sodium at the recommended rate has no adverse effect on soil structure even on soils of low structural stability.

If inorganic fertilisers containing potash and sodium are applied just before sowing and too close to the seed, plant populations can be reduced in dry conditions, especially on sandy soils. To minimise this risk, all inorganic fertilisers should be applied at least two weeks before sowing and incorporated into the soil. They may be applied in autumn or winter and ploughed in except on light sand soils where there is a risk of some nutrient loss by leaching. On the latter soils, the fertilisers can be applied in January/February just before ploughing or cultivating.

Boron

Boron deficiency can adversely affect sugar beet yields. An application of boron may be required where soil analysis indicates that available boron in the soil (hot water extraction) is less than 0.8 mg B/kg (ppm B). Deficiency can be corrected by applying 3 kg B/ha. Seek advice from a FACTS Qualified Adviser about form, amount and timing of the application.

Don't forget to:

make an allowance for nutrients applied in organic manures (see Section 2)

ensure the phosphate and potash offtake is balanced by application on Index 2 soils and check that the soil is maintained at Index 2 by soil sampling every 3 – 5 years

Section 4: Arable and forage crops

Forage Maize

The amounts of phosphate and potash shown at target Index 2 are needed to replace the offtakes in a fresh yield of 40 t/ha, and to maintain the soil at the target Index. The phosphate and potash recommendations at target or lower Indices can be adjusted if yields are likely to be larger or smaller than 40 t/ha by multiplying the difference in expected yield by the phosphate and potash content per tonne of yield given in Appendix 5. For example, at P Index 1, the recommendation for an expected yield of 50 t/ha is $85 + (10 \times 1.4) = 99$ kg P_2O_5/ha. No adjustment for yield should be made where the soil Index is higher than target.

Crops grown on soil at Index 0 and 1 would be expected to respond to the extra amounts of phosphate, potash and magnesium shown in the table below. Also over a number of years, these extra amounts of fertiliser will help to raise the Index level of most soils, except light sands.

	SNS, P or K Index				
	0	1	2	3	4 and higher
	kg N/ha				
Nitrogen (N) All mineral soils	150	100	50	20	0
Phosphate (P_2O_5)	115	85	55	20	0
Potash (K_2O)	235	205	175 (2-) 145 (2+)	110	0

Using organic manures in rotations that include maize

Large quantities of organic manures are commonly applied before maize is grown. **The recommendations in the table apply to crops where organic manures are not applied.** The nutrients supplied from recent and past applications of organic manures must be allowed for when deciding on fertiliser application rates (see Section 2).

In an NVZ, applications of manufactured nitrogen fertiliser and organic manures must comply with all of the relevant NVZ Action Programme measures. On land that is not in an NVZ, organic manure applications should follow the recommendations in *Protecting Our Water, Soil and Air: A Code of Good Agricultural Practice* (see Section 9), which includes a limit of 250 kg N/ha of organic manure total nitrogen every 12 months.

Where maize is grown continuously and organic manures are also regularly used, excessively large amounts of nitrogen and phosphate can build up in the soil. This can greatly increase the risk of nitrate and phosphate transfer to watercourses. *Protecting Our Water, Soil and Air: A Code of Good Agricultural Practice* states that, where the soil P Index is 3 and organic manures are used, no more than maintenance rates of phosphate should be applied during the rotation.

To minimise the risk of building up a large excess of nutrient in soil, maize should be grown in rotation with other crops so that an acceptable nutrient balance is reached for the rotation. In maize rotations, extra care must be given to avoiding excessive nutrient inputs as organic manures or fertilisers.

Section 4: Arable and forage crops

Placement of nitrogen and phosphate

To encourage rapid early growth, all of the phosphate requirement and up to 10-15 kg/ha of the nitrogen requirement may be placed below the seed at drilling. The remainder of the nitrogen requirement should be top-dressed as soon as the crop has emerged.

Potash

Potash should be applied before seedbed preparation and thoroughly worked in.

Magnesium

Where sugar beet or potatoes do not feature in the rotation, magnesium fertiliser is only justified at soil Index 0 when 50 to 100 kg MgO/ha should be applied every three or four years (see page 42).

Don't forget to:

make allowance for nutrients applied in organic manures (see Section 2)

ensure the phosphate and potash offtake is balanced by application on Index 2 soils and check that the soil is maintained at Index 2 by soil sampling every 3 – 5 years

Section 4: Arable and forage crops

Other Forage Crops

The amounts of phosphate and potash shown at target Index 2 are needed to replace the offtakes in the fresh yields of each crop as shown in the table, and to maintain the soil at the target Index. The phosphate and potash recommendations at target or lower Indices can be adjusted if yields are likely to be larger or smaller than those shown by multiplying the difference in expected yield by the phosphate and potash content per tonne of yield given in Appendix 5. For example, at P Index 1, the recommendation for forage swedes with an expected yield of 75 t/ha is 75 +(10 x 0.7) = 82 kg P_2O_5/ha. No adjustment for yield should be made where the soil Index is higher than target.

Crops grown on soil at Index 0 and 1 would be expected to respond to the extra amounts of phosphate, potash and magnesium shown in the table below. Also over a number of years, these extra amounts of fertiliser will help to raise the Index level of most soils, except light sands.

	SNS Index						
	0	1	2	3	4	5	6
	kg N/ha						
Forage Swedes and Turnips (65 t/ha roots removed)							
Nitrogen (N)	100	80	60	40	0-40	0	0
Phosphate (P_2O_5)	105	75	45	0	0	0	0
Potash (K_2O)	215	185	155 (2-) 125 (2+)	80	0	0	0
Forage Rape and Stubble Turnips (grazed)							
Nitrogen (N)	100	90	80	60	40	0-40	0
Phosphate (P_2O_5)	85	55	25	0	0	0	0
Potash (K_2O)	110	80	50 (2-) 20 (2+)	0	0	0	0
Fodder Beet and Mangels (65 t/ha roots removed)							
Nitrogen (N)	130	120	110	90	60	0-40	0
Phosphate (P_2O_5)	110	80	50	0	0	0	0
Potash (K_2O)	170	140	110 (2-) 80 (2+)	40	0	0	0
Kale (40 t/ha cut)							
Nitrogen (N)	130	120	110	90	60	0-40	0
Phosphate (P_2O_5)	110	80	50	0	0	0	0
Potash (K_2O)	260	230	200 (2-) 170 (2+)	130	0	0	0
Forage Rye and Forage Triticale (20 t/ha cut)							
Nitrogen (N)	80	60	40	20	0	0	0
Phosphate (P_2O_5)	95	65	35	0	0	0	0
Potash (K_2O)	180	150	120 (2-) 90 (2+)	50	0	0	0

Section 4: Arable and forage crops

Forage rape and stubble turnips

When grown as a catch crop after cereals, apply no more than 75 kg N/ha at Index 0 or 1. Further reductions may be made if the soil is moist and has been cultivated. For stubble turnips sown after mid August, apply 50 kg P_2O_5/ha at Index 0 only.

Fodder beet and mangels

Salt is recommended on all soils except Fen silts and peats. Apply 400 kg/ha of agricultural salt (200 kg Na_2O /ha) well before drilling. If sodium is recommended but not applied, increase potash by 100 kg K_2O/ha.

A boron application may be needed. Soil analysis is a useful guide to assess the need for boron.

Phosphate and potash

The recommendations assume that crops are grazed. Where tops are carted off, potash applications may need to be increased by up to 150 kg K_2O/ha. Phosphate and potash need only be applied to the seedbed at Index 0 or 1.

Magnesium

Where sugar beet or potatoes do not feature in the rotation, magnesium fertiliser is only justified at soil Index 0 when 50 to 100 kg MgO /ha should be applied every three or four years (see page 42).

Don't forget to:

make allowance for nutrients applied in organic manures (see Section 2)

ensure the phosphate and potash offtake is balanced by application on Index 2 soils and check that the soil is maintained at Index 2 by soil sampling every 3 – 5 years

Section 4: Arable and forage crops

Ryegrass Grown for Seed

	SNS, P or K Index						
	0	1	2	3	4	5	6
	kg N/ha						
Nitrogen (N)							
Light sand soils	160	110	60	0-40	0	0	0
All other mineral soils		160	110	60	0-40	0	0
Organic soils				60	0-40	0	0
Peaty soils						0-40	
Phosphate (P$_2$O$_5$)	90	60	30	0	0	0	0
Potash (K$_2$O)	150	120	90 (2-) 60 (2+)	0	0	0	0

Nitrogen

Nitrogen rates are for crops where there is a low risk of crop lodging either due to field characteristics or use of a growth regulator. A lower nitrogen rate will be appropriate for crops with a higher risk of lodging. Higher rates may be needed in the second cropping year or where amenity varieties are grown.

Where the recommended rate is 100 kg N/ha or more, apply 40 kg N/ha in early-mid March and the balance of the application in early April. Where the requirement is less than 100 kg N/ha, apply the whole application in early April.

Where a crop of Italian ryegrass seed is to be grown following a silage crop, apply 60 kg N/ha immediately following the silage crop.

Phosphate and potash

The amounts of phosphate and potash shown at target Index 2 are needed to replace the offtakes and maintain the soil at the target Index. Crops grown on soil at Index 0 and 1 would be expected to respond to the extra amounts of phosphate, potash and magnesium shown in the table below. Also over a number of years, these extra amounts of fertiliser will help to raise the Index level of most soils, except light sands.

Phosphate and potash can be applied at any convenient time except at Index 0 and 1 when the dressing should be applied in the spring of the harvest year.

Don't forget to:

make allowance for nutrients applied in organic manures (see Section 2)

ensure the phosphate and potash offtake is balanced by application on Index 2 soils and check that the soil is maintained at Index 2 by soil sampling every 3 – 5 years

Section 5: Vegetables and Bulbs

	Pages
Checklist for decision making	134
Fertiliser use for vegetables	135
Asparagus	138
Brussels sprouts and cabbages	139
Cauliflowers and calabrese	141
Self blanching celery	143
Peas and beans	144
Lettuce, radish, sweetcorn and courgettes	145
Onions and leeks	147
Root vegetables	149
Bulbs and bulb flowers	151
Calculation of the crop nitrogen requirement (CNR)	Appendix 9

This section provides guidance for the application of fertilisers to field vegetable crops. Before looking at the individual crop recommendations refer to the 'Checklist for decision making' and 'Fertiliser use for vegetables'. Vegetables have specific requirements which must be met to ensure efficient production.

More background on the principles underlying the recommendations is given in Section 1 'Principles of Nutrient Management and Fertiliser Use'.

For each crop listed in the table above, recommendations for nitrogen (N), phosphate (P_2O_5) and potash (K_2O) are given in kg/ha. Sulphur (SO_3), magnesium (MgO) and sodium (Na_2O) recommendations, also in kg/ha, are given where these nutrients are needed. Recommendations are given for the rate and timing of fertiliser application.

The recommendations are based on the nutrient requirements of the crops being grown, making allowance for the nutrients already present in the soil. Section 3 gives instructions on how to use the tables, including how to determine the SNS Index.

June 2010

Section 5: Vegetables and Bulbs

Checklist for Decision Making

Decisions for fertiliser use must be made for each field separately. Where more than one crop is grown in a field, crops must be considered individually.

1. Confirm the crop to be grown and the intended market. Identify any specific crop quality requirements for this market.

2. Identify the dominant soil type in the cropped area (see page 86 and Appendix 1).

3. Carry out soil analysis for pH, P, K and Mg every 3-5 years (see page 35 and Appendix 3). Target values for vegetable rotations are:

   Soil pH 6.5 or 7.0 for brassicae if clubroot is a problem (pH 5.8 on peat soils))
   Soil P Index 3, K Index 2+, Mg Index 2

4. Identify the SNS Index of the field either by carrying out soil analysis for soil mineral nitrogen (the Measurement Method, page 95) or by the Field Assessment Method based on previous cropping, previous fertiliser and manure use, soil type and winter rainfall (see page 86). SMN analysis is recommended where nitrogen residues are expected to be moderate or high (e.g. following previous manure use, crops receiving high rates of nitrogen fertiliser or for the second crop in one season). Special care is needed when interpreting soil nitrogen analysis data for shallow rooted vegetable crops or where the rooting depth is restricted (Section 3).

5. Calculate the nutrients that will be available for crop uptake from organic manures and green waste that have been applied since harvest of the previous crop (Section 2). If using the Field Assessment Method, deduct these nutrients from the recommended rates given in the tables. (Crop assurance schemes and protocols may restrict manure application).

6. Decide on the strategy for phosphate and potash use. This will be either building up, maintaining or running down the soil Index levels (see pages 38-41). Allow for any surplus or deficit of phosphate or potash applied to previous crops in the rotation.

7. Calculate the amount of phosphate and potash removed in the harvested crop (see pages 38-41 and Appendix 5). This is the amount of these nutrients that must be replaced in order to maintain the soil at the target Index.

8. Decide if starter fertiliser or banded fertiliser would be appropriate (see 'Fertiliser Use for Vegetables', below).

9. Using the tables, decide on the required rate of each nutrient. Decide on the optimum timings for fertiliser application; then find the best match for these applications using available fertilisers (see page 102).

10. Check that the fertiliser spreader is in good working order and has been recently calibrated (see page 49).

11. Keep an accurate record of all fertilisers and organic manures applied.

Section 5: Vegetables and Bulbs

Fertiliser Use for Vegetables

The principles governing the use of fertilisers are given in Section 1.

Crop nitrogen requirement

Where sufficient data are available the nitrogen recommendations are based on a three-step process:

- Size of the crop – the size, frame, or weight of the crop needed to produce optimal economic yields.
- Nitrogen uptake – the optimum nitrogen uptake associated with a crop of that size.
- Supply of nitrogen – based on the nitrogen supply from the soil within rooting depth including any nitrogen mineralised from organic matter during the growing season.

Recommendations are given for typical crops produced in the main part of the growing season. As vegetable crops are planted at many different times of the year and have a range of expected yields, the table in Appendix 10 can be used to customise individual field recommendations. Earlier planted crops may need extra nitrogen because the supply of nitrogen from mineralisation is less than later in the growing season. Some vegetable such as beetroot can have wide ranging yield potential depending on the market. The baby beet crop will have a smaller nitrogen demand but is shallow rooted compared with larger yielding processing crops so will require similar amounts of nitrogen.

The recommendations assume effective pest and disease control. Where crops are grown with minimal control measures or the crop is intended for storage, smaller amounts of nitrogen fertiliser should be considered. In all cases too much nitrogen fertiliser can give rise to poor quality crops especially when growing conditions are difficult. Where large amounts of nitrogen residues from previous crops are expected measurement of soil mineral nitrogen can be helpful.

The use of a decision support system, such as WELL_N to help interpretation of soil nitrogen measurements can be advantageous. Decision support systems can also assist in assessing the need to vary nitrogen recommendations particularly for high or low yielding crops or in cases of extreme weather where nitrogen may have been leached out of the immediate rooting depth of a crop.

Nitrogen residues following vegetable crops

On deep silty or deep clayey soils, nitrogen residues in predominantly vegetable rotations can persist for several years especially in the drier areas. These residues are likely to be evident following 'High or medium N vegetables'. The SNS tables make some allowance for this long persistency of nitrogen residues. However, the Index may need to be adjusted upwards particularly where winter rainfall is low or where the history of vegetable cropping is longer than one year, and in circumstances where larger than average amounts of crop residue or unused fertiliser remain in the soil to rooting depth (see Tables A-C in Section 2). In rotations where vegetable crops are grown infrequently in essentially arable rotations, the Index may need to be adjusted downwards.

Section 5: Vegetables and Bulbs

Nitrogen rich leafy trash from many brassica crops is ploughed into the land at various times of the year. The nitrogen in these materials can become available for use by the next crop very rapidly in summer but more slowly in the winter when the soil temperature is lower. In this situation where double cropping is practised in the summer season, the SNS Index can be increased by 1 Index if following 'Medium N vegetables' and by up to 2 following 'High N vegetables' as indicated in Section 2. It is important that the growing conditions of the first crop are fully taken into account, especially where nitrogen may be leached from light sand soils in wet seasons or where excess irrigation has been applied. Sampling and analysis for SMN before the second crop is worthwhile.

Care needs to be taken where residues are ploughed in after late December; the nitrogen may not become available for uptake by the next crop until after that crop requires the bulk of its nitrogen supply.

Where there is uncertainty about the amount of nitrogen in the soil, soil sampling for Soil Mineral Nitrogen (SMN) may be appropriate.

Assessing Soil Nitrogen Supply (SNS)

Many vegetable crops are shallow rooted and cannot take full advantage of mineral nitrogen to 90 cm depth. The SNS Index tables may indicate relatively high levels of SNS from previous crops, but not all of this nitrogen will be in the rooting zone of shallow rooted crops. This will be particularly so following wet winters on medium soils where nitrogen is leached to lower depths. In this situation, even deep-rooted crops may require small dressings of nitrogen (up to 50 kg N/ha) to support establishment. Soil mineral nitrogen sampling is recommended to identify the distribution of mineral nitrogen with depth.

Soil sampling should be carried out to the rooting depth specified in Appendix 2. The amount of nitrogen to rooting depth can be used to estimate soil mineral nitrogen to 90 cm assuming that the distribution of mineral nitrogen is uniform. i.e. 0-90 = 0-60 cm value x 1.5. The recommendation tables will take account only of nitrogen available to the depth sampled.

Phosphate and potash

Many vegetables respond to fresh applications of phosphate and potash fertiliser especially at low soil indices, it is important that these needs are fully met. At low soil indices, there is the choice to build these levels up to the target Index. For this purpose extra amounts of phosphate and potash can be applied on Index 0 and 1 soils as shown in the tables below, the amounts vary according to the responsiveness of the crop grown. The amount needed to supply maintenance needs will depend on the yields of the crops expected and the treatment of crop residues. For a more precise calculation of maintenance requirements Appendix 5 contains information on nutrient contents of marketable product removed from the field.

Additionally there are instances where small amounts of nitrogen and phosphate fertiliser placed beneath seedlings or transplants can improve establishment, early growth and subsequent use of nutrients. The use of these techniques is encouraged but the amount in any starter dose applied should be deducted from the total base application required.

Section 5: Vegetables and Bulbs

Sulphur

Many field vegetable crops, particularly brassicas have a significant requirement for sulphur. Experience in the arable sector has shown useful responses to sulphur on a third of the cereal acreage. Whilst there have been few UK-based trials, there is evidence that brassica crops do respond to sulphur. In situations where sulphur levels might be low following wet winters on light soils with no previous history of manure, use of sulphur-containing fertilisers should be considered as a base dressing to supply nitrogen and sulphur.

Techniques for applying fertiliser

Starter fertiliser

The injection of high phosphate liquid fertiliser 2-3 cm below the seed, or around the roots of transplants, can improve the growth and quality of crops such as bulb and salad onions, lettuce and leeks. Starter fertiliser is particularly useful for crops grown in mixed rotations on soils at P Index 3 or below. However, responses have been found at P Index 4. No more than 20 kg N/ha and 60 kg P_2O_5/ha should be applied as starter fertiliser which may be deducted from the recommended total application. In most experiments comparable yields with starter fertiliser have been obtained with much lower amounts of nitrogen than when fertiliser has been broadcast.

Liquid starter fertilisers containing chloride may reduce plant establishment and should be avoided.

Band spreading/placement of nitrogen

For some wide row crops there can be benefits from applying early nitrogen in a band or injecting it around the plant, followed by a broadcast top-dressing. This may reduce the overall amount of nitrogen required. In experiments using the banding approach, yield and quality of cauliflowers was maintained. Post harvest soil nitrogen residues were less, reducing the risk of nitrate leaching.

Fertigation

Fertigation, applying nutrients in irrigation water, has been shown to produce batavia lettuce with better quality compared with broadcasting fertiliser because nutrients and water can be more effectively targeted to crop need. Experiments have demonstrated that by using fertigation, savings of up to 33% in nitrogen applications can be made. Whilst the technique has the potential to reduce nitrogen use – some preliminary testing should be carried out to fine-tune the amounts and timing of nutrients especially on crops other than lettuce.

Nitrification inhibitors

These products slow the conversion of ammonium-N to nitrate-N. They are available in manufactured solid and liquid fertilisers and can be added to liquid fertilisers prior to application or sprayed onto soil prior to spreading solid fertilisers. Nitrification inhibitors can delay release of nitrate following fertiliser application which can reduce nitrate leaching and nitrous oxide emissions.

Section 5: Vegetables and Bulbs

Asparagus

	SNS, P, K or Mg Index					
	0	1	2	3	4	5 or higher
	kg/ha					
Establishment year						
Nitrogen(N) – all soil types	150	150	150	90	20	0
Phosphate (P_2O_5)	175	150	125	100	75	0
Potash (K_2O)	250	225	200	150	125	0
Subsequent years						
Nitrogen (N) – year 2, all soil types	see note below					
Nitrogen (N) – other years, all soil types	see note below					
Phosphate (P_2O_5)	75	75	50	50	25	0
Potash (K_2O)	100	50	50	50	0	0
Magnesium (MgO)	150	100	0	0	0	0

Establishment year – nitrogen

Apply one third of the total nitrogen dressing before sowing or planting, one third when the crop is fully established (around mid June for crowns, mid July for transplants) and one third at the end of August.

Subsequent years – nitrogen

In year 2, apply 120 kg N/ha by end-February – early March.

In subsequent years, the amount and timing of nitrogen depends on the previous winter. If the crop is on light soil and over-winter rainfall was high, apply 40 – 80 kg N/ha by the end of February with an additional 40-80 kg N/ha applied after harvest.

Following moderate or low amounts of winter rainfall apply 40-80 kg N/ha just after the harvest to provide nitrogen for fern growth.

Where SMN is measured, top up with fertiliser nitrogen to achieve a target of 120 kg N/ha of mineral nitrogen in the top 30 cm of soil during the cropping period.

Sodium

Asparagus can respond to applied sodium. Apply up to 500 kg Na_2O /ha per year at the end of June but not in the establishment year.

Don't forget to:

make allowance for nutrients applied in organic manures (see Section 2)

ensure the phosphate and potash offtake is balanced by application on Index 2 soils and check that the soil is maintained at Index 2 by soil sampling every 3 – 5 years

Section 5: Vegetables and Bulbs

Brussels Sprouts and Cabbage

	SNS, P, K or Mg Index						
	0	1	2	3	4	5	6
	kg/ha						
Nitrogen (N)[b] – all soil types							
Brussels sprouts	330	300	270	230	180	80	0[a]
Storage cabbage	340	310	280	240	190	90	0[a]
Head cabbage pre-December 31st	325	290	260	220	170	70	0[a]
Head cabbage post-December 31st	240	210	180	140	90	0[a]	0[a]
Collards – pre- December 31st	210	190	180	160	140	90	0[a]
Collards – post- December 31st	310	290	270	240	210	140	90
Phosphate[c] (P_2O_5), all crops	200	150	100	50	0	0	0
Potash[c] (K_2O), all crops	300	250	200 (2-) 150 (2+)	60	0	0	0
Magnesium (MgO), all crops	150	100	0	0	0	0	0

a. A small amount of nitrogen may be needed if there is little mineral nitrogen in the 0-30 cm of soil.

b. Nitrogen – On light soils where leaching may occur or when crops are established by direct seeding no more than 100 kg N/ha should be applied prior to seeding or transplanting. On retentive soils in drier parts of the country where leaching risk is low and spring planted brassicas are established from modules, more nitrogen can be applied prior to planting. The remainder of the nitrogen requirement should be applied after establishment but before the surface soil dries out to ensure that it is utilised by the crop.

c. Phosphate and potash requirements are for average crops and it is important to calculate specific phosphate and potash removals based on yields especially for the larger yielding cabbage crops (see 'Fertiliser Use for Vegetables' above and Section 1).

Storage cabbage

For storage cabbage grown on fertile soils the recommendations for nitrogen may need to be decreased in order to reduce the risks of storage losses.

Post-December 31st crops

Apply no more than 100 kg N/ha at sowing or transplanting, less if there is risk of frost damage. The remaining nitrogen should be applied to reflect crop growth. Further top dressings of nitrogen will depend on the harvest date and expected yield – some nitrogen will be required to support growth during the winter particularly for crops harvested in late winter. For crops harvested in late spring more of the top-dressing should be left until the beginning of regrowth in spring.

Section 5: Vegetables and Bulbs

Sulphur

Consider applying up to 50 kg SO_3/ha in situations where sulphur content of soils is low, i.e on light soils following wet winters where there is no history of organic manure application.

Don't forget to:

make allowance for nutrients applied in organic manures (see Section 2)

ensure the phosphate and potash offtake is balanced by application on Index 2 soils and check that the soil is maintained at Index 2 by soil sampling every 3 – 5 years

Section 5: Vegetables and Bulbs

Cauliflowers and Calabrese

	SNS, P, K or Mg Index						
	0	1	2	3	4	5	6
	kg/ha						
Nitrogen (N) – all soil types							
Cauliflower, summer/autumn[a]	290	260	235	210	170	80	0[b]
Cauliflower, winter hardy/roscoff[a]							
– seedbed	100	100	100	100	60	0[a]	0[b]
– top-dressing	190	160	135	110	100	80	0[b]
Calabrese[a]	235	200	165	135	80	0[b]	0[b]
Phosphate (P_2O_5)	200	150	100	50	0	0	0
Potash (K_2O)	275	225	175 (2-) / 125 (2+)	35	0	0	0
Magnesium (MgO)	150	100	0	0	0	0	0

a. The recommendations assume overall application. Band spreading of nitrogen may be beneficial (see 'Techniques for Applying Fertiliser' above).

b. A small amount of nitrogen may be needed if soil nitrogen levels are low in the 0-30 cm of soil (see 'Techniques for Applying Fertiliser' above).

Cauliflower and calabrese

Where there is a risk of poor establishment or leaching, apply no more than 100 kg N/ha at sowing or transplanting. The remainder should be applied when the crop is established but before the surface soil dries out.

There is a benefit from banding or placing the nitrogen applied at sowing or transplanting – if nitrogen is only applied to half the width of the row reduce the seedbed application by 33%.

The SNS Index for second crops grown in the same season is likely to be between Index 4 and 6 depending on the growing conditions of the first crop (Section 2)

Cauliflower, winter hardy/roscoff

Apply no more than 100 kg N/ha at sowing or transplanting, less if there is risk of frost damage. The amount of nitrogen applied subsequently will depend on crop growth, for example up to 60 kg N/ha per month in the south west and 20 kg N/ha in the north.

Where seedbed SNS exceeds 4 and crops are likely to be harvested in April or later the top-dressing should be left until the start of growth in the spring but then the SNS may need to be recalculated to take account of any over-winter losses of nitrogen, uptake of nitrogen by the crop as well as mineral nitrogen to 90 cm.

Section 5: Vegetables and Bulbs

Sulphur

Consider applying up to 50 kg SO_3/ha in situations where sulphur content of soils is low, for example on light soils following wet winters where there is no history of organic manures application.

Don't forget to:

make allowance for nutrients applied in organic manures (see Section 2)

ensure the phosphate and potash offtake is balanced by application on Index 2 soils and check that the soil is maintained at Index 2 by soil sampling every 3-5 years

Section 5: Vegetables and Bulbs

Self-Blanching Celery

	SNS, P, K or Mg Index						
	0	1	2	3	4	5	6
	kg/ha						
Nitrogen (N) – all soil types							
– seedbed	75	75	75	75	0[a]	0[a]	0[a]
– top-dressing[b]	see note below table						
Phosphate (P_2O_5)	250	200	150	100	50	0	0
Potash (K_2O)	450	400	350 (2-) 300 (2+)	210	50	0	0
Magnesium (MgO)	150	100	0	0	0	0	0

a. A small amount of nitrogen may be needed if soil nitrogen levels are low in the 0-30 cm of soil.

b. A top-dressing of 75-150 kg N/ha will be required 4-6 weeks after planting.

Sodium

Celery is responsive to sodium which is recommended for celery grown on all soils except peaty and some Fen silt soils, which generally contain adequate amounts of sodium. Sodium can be applied as agricultural salt at 375 kg/ha (200 kg Na_2O/ha). The application will not have any adverse effect on soil structure, even on soils of low structural stability.

Don't forget to:

make allowance for nutrients applied in organic manures (see Section 2)

ensure the phosphate and potash offtake is balanced by application on Index 2 soils and check that the soil is maintained at Index 2 by soil sampling every 3 – 5 years

Section 5: Vegetables and Bulbs

Peas (Market Pick) and Beans

	SNS, P, K or Mg Index						
	0	1	2	3	4	5	6
	kg/ha						
Peas for Fresh Market							
Nitrogen(N) – all soil types	0	0	0	0	0	0	0
Phosphate (P_2O_5)	185	135	85	35	0	0	0
Potash (K_2O)	190	140	90 (2-) 40 (2+)	0	0	0	0
Magnesium (MgO)	100	50	0	0	0	0	0
Beans							
Nitrogen(N) – all soil types							
Broad beans	0	0	0	0	0	0	0
Dwarf and runner beans – seedbed	180	150	120	80	30	0[a]	0[a]
Runner beans – top-dressing	see note below						
Phosphate (P_2O_5)	200	150	100	50	0	0	0
Potash (K_2O)	200	150	100 (2-) 50 (2+)	0	0	0	0
Magnesium (MgO)	100	50	0	0	0	0	0

a. A small amount of nitrogen may be needed if soil nitrogen levels are low in the 0-30 cm of soil.

Peas for fresh market

Peas may suffer from sulphur deficiency on sensitive soil types (page 43). Where deficiency is possible, apply 25 kg SO_3/ha.

Dwarf/Runner beans

Apply no more than 100 kg N/ha at sowing or planting. The remainder should be applied when the crop is fully established.

Runner beans can require a further top-dressing of up to 75 kg N/ha at early picking stage.

Don't forget to:

make allowance for nutrients applied in organic manures (see Section 2)

ensure the phosphate and potash offtake is balanced by application on Index 2 soils and check that the soil is maintained at Index 2 by soil sampling every 3 – 5 years

Section 5: Vegetables and Bulbs

Lettuce, Radish, Sweetcorn and Courgettes

	SNS, P, K or Mg Index						
	0	1	2	3	4	5	6
	kg/ha						
Lettuce							
Nitrogen(N) – all soil types	200	180	160	150	125	75	30
Phosphate (P_2O_5)[a]	250	200	150	100	*	*	0
Potash (K_2O)	250	200	150 (2-) 100 (2+)	0	0	0	0
Magnesium (MgO)	150	100	0	0	0	0	0
Radish, Sweetcorn and Courgettes							
Nitrogen(N) – all soil types							
Radish	100	90	80	65	50	20	0[b]
Sweet Corn	150	100	50	0[b]	0[b]	0[b]	0[b]
Courgettes – seedbed	100	100	100	40	0[b]	0[b]	0[b]
– top-dressing	see note below						
Phosphate (P_2O_5)	175	125	75	25	0	0	0
Potash (K_2O)	250	200	150 (2-) 100 (2+)	0	0	0	0
Magnesium (MgO)	150	100	0	0	0	0	0

a. The recommendations assume overall application. A starter fertiliser containing nitrogen and phosphate may be beneficial (see 'Techniques for Applying Fertiliser' above).

b. A small amount of nitrogen may be needed if soil nitrogen levels are low in the 0-30 cm of soil (see 'Techniques for Applying Fertiliser' above).

* At P Index 4 and 5, phosphate up to 60 kg P_2O_5/ha as starter fertiliser may be beneficial (see 'Techniques for Applying Fertiliser' above).

Lettuce

These recommendations are provided for the larger Crisp and Escarole type lettuce, while other lower yielding types such as Lollo Rossa, Little Gem, Cos, Endives and Butterhead may need less nitrogen. Each situation will need to be judged carefully as rooting depth of the lower yielding crops is likely to be only 30 cm, compared with 60 cm for the larger crop so less of the soil nitrogen will be available.

Apply no more than 100 kg N/ha at sowing or planting on light sandy soils. The remainder should be applied when the crop is fully established. When crop covers are used all the nitrogen will need to be applied as a base dressing but care should be taken to avoid poor establishment in dry soils.

Starter fertilisers containing nitrogen and phosphate can provide equivalent crop yields, with lower amounts of nitrogen than from broadcast fertiliser.

Section 5: Vegetables and Bulbs

Fertigation can provide crops of better quality as nutrients and water can be more effectively targeted to crop need. Experiments have demonstrated savings of up to 33% in nitrogen applications compared with broadcast fertilisers.

Where more than one crop is grown in the same year there should be sufficient residues of phosphate, potash and magnesium for a second crop. The SNS Index for second crops grown in the same season will be between Index 3 and 5 depending on the growing conditions of the first crop (see 'Fertiliser Use for Vegetables' above).

Radish and sweet corn

Apply no more than 100 kg N/ha in the seedbed. Apply the remainder as a top-dressing when the crop is fully established.

Courgettes

Top-dressings of up to 75 kg N/ha in total may be required.

Don't forget to:

make allowance for nutrients applied in organic manures (see Section 2)

ensure the phosphate and potash offtake is balanced by application on Index 2 soils and check that the soil is maintained at Index 2 by soil sampling every 3 – 5 years

Section 5: Vegetables and Bulbs

Onions and Leeks

	\multicolumn{7}{c}{SNS, P, K or Mg Index}
	0	1	2	3	4	5	6
	\multicolumn{7}{c}{kg/ha}
Nitrogen(N) – all soil types							
– Bulb Onions[a]	160	130	110	90	60	0[b]	0[b]
– Salad Onions[a]	130	120	110	100	80	50	20
– Leeks[a]	200	190	170	160	130	80	40
Phosphate (P_2O_5)	200	150	100	50	*	*	0
Potash (K_2O)	275	225	175 (2-) 125 (2+)	35	0	0	0
Magnesium (MgO)	150	100	0	0	0	0	0

a. The recommendations assume overall application. A starter fertiliser containing nitrogen and phosphate may be beneficial (see 'Techniques for Applying Fertiliser' above).

b. A small amount of nitrogen may be needed if soil nitrogen levels are low in the 0-30 cm of soil (see 'Techniques for Applying Fertiliser' above).

* At P Index 4 and 5, phosphate up to 60 kg P_2O_5/ha fertiliser as starter fertiliser may be justified (see 'Techniques for Applying Fertiliser' above).

Bulb onions

At SNS Index 0 on light sands where spring soil mineral nitrogen levels are 40 kg N/ha or less a further 15 kg N/ha can be supplied.

Apply no more than 100 kg N/ha to the seedbed. The remainder should be applied when the crop is fully established for the spring crop and the following spring for the autumn sown crop.

Salad onions

At SNS Index 0 on light sands where spring soil mineral nitrogen levels are 40 kg N/ha or less a further 15 kg N/ha can be supplied.

Apply no more than 100 kg N/ha to the seedbed of the spring sown crop. The remainder should be applied when the crop is fully established.

For the autumn sown crop care must be taken not to apply too much nitrogen as the crop is prone to disease. Apply not more than 40 kg N/ha. If the crop is planted on organic or peaty soils or where large amounts of crop residue have been incorporated no seedbed nitrogen is required. The remainder should be applied the following spring.

Section 5: Vegetables and Bulbs

Leeks

Apply no more than 100 kg N/ha in the seedbed. The remainder should be applied as a top-dressing when the crop is fully established. An additional top-dressing of 100 kg N/ha may be required on all soils except peat depending on the appearance of the crop, to support growth and colour.

Don't forget to:

make allowance for nutrients applied in organic manures (see Section 2)

ensure the phosphate and potash offtake is balanced by application on Index 2 soils and check that the soil is maintained at Index 2 by soil sampling every 3 – 5 years

Section 5: Vegetables and Bulbs

Root Vegetables

	SNS, P, K or Mg Index						
	0	1	2	3	4	5	6
	kg/ha						
Beetroot							
Nitrogen(N) – all soil types	290	260	240	220	190	120	60
Phosphate (P_2O_5)	200	150	100	50	0	0	0
Potash (K_2O)	300	250	200 (2-) 150 (2+)	60	0	0	0
Swedes							
Nitrogen(N) – all soil types	135	100	70	30	0[a]	0[a]	0[a]
Phosphate (P_2O_5)	200	150	100	50	0	0	0
Potash (K_2O)	300	250	200 (2-) 150 (2+)	60	0	0	0
Turnips, Parsnips							
Nitrogen(N) – all soil types	170	130	100	70	20	0[a]	0[a]
Phosphate (P_2O_5)	200	150	100	50	0	0	0
Potash (K_2O)	300	250	200 (2-) 150 (2+)	0	0	0	0
Carrots							
Nitrogen(N) – all soil types	100	70	40	0[a]	0[a]	0[a]	0[a]
Phosphate (P_2O_5)	200	150	100	50	0	0	0
Potash (K_2O)	275	225	175 (2-) 125 (2+)	35	0	0	0
All Crops							
Magnesium (MgO)	150	100					

a. A small amount of nitrogen may be needed if soil nitrogen levels are low in the 0-30 cm of soil (see 'Fertiliser use for vegetables' above).

Nitrogen – all crops

Apply no more than 100 kg N/ha in the seedbed. The remainder should be applied as a top-dressing when the crop is fully established.

Section 5: Vegetables and Bulbs

Phosphate and potash – all crops

High yielding root crops can take up large amounts of phosphate and potash. The amounts removed can be calculated from the known yield and the amount of phosphate and potash per tonne fresh produce shown in Appendix 5. It is important to do this to maintain the target Index for both phosphate and potash. Where straw is used to protect carrots and is subsequently incorporated into the soil, it contributes approximately 1 and 8 kg phosphate and potash, respectively, per tonne of straw (see Appendix 5). These amounts should be considered when calculating the phosphate and potash requirements of following crops.

Carrots – sodium

On sandy soils apply 200 kg Na_2O/ha as salt and deeply cultivate the application into the soil before drilling.

Don't forget to:

make allowance for nutrients applied in organic manures (see Section 2)

ensure the phosphate and potash offtake is balanced by application on Index 2 soils and check that the soil is maintained at Index 2 by soil sampling every 3 – 5 years

Section 5: Vegetables and Bulbs

Bulbs and Bulb Flowers

	SNS, P, K or Mg Index						
	0	1	2	3	4	5	6
	kg/ha						
Nitrogen(N) – all soil types	125	100	50	0	0	0	0
Phosphate (P_2O_5)	200	150	100	50	0	0	0
Potash (K_2O)	300	250	200 (2-) 150 (2+)	60	0	0	0
Magnesium (MgO)	150	100	0	0	0	0	0

Timing of application

Apply nitrogen as a top-dressing just before emergence.

Narcissus, subsequent years

A top-dressing of 50 kg N/ha may be required in the second or subsequent year if growth was poor in the previous year.

Don't forget to:

make allowance for nutrients applied in organic manures (see Section 2)

ensure the phosphate and potash offtake is balanced by application on Index 2 soils and check that the soil is maintained at Index 2 by soil sampling every 3 – 5 years

Section 6: Fruit, Vines and Hops

	Page
Checklist for decision making	154
Fertiliser use for fruit, vines and hops	155
Fruit, vines and hops – before planting	157
Top fruit – established orchards	159
Soft fruit and vines – established plantations	162
Leaf analysis for top and soft fruit	166
Apple fruit analysis	168
Hops	171

For each crop, recommendations for nitrogen (N), phosphate (P_2O_5), potash (K_2O) and magnesium (MgO) are in kg/ha. Recommendations are given for the rate and timing of fertiliser application.

The following recommendations are based on the nutrient requirements of the crop being grown, making allowance for the nutrients already in the soil. Section 3 gives instructions on how to use the tables, including assessment of the SNS Index.

More background on the principles underlying the recommendations is given in Section 1.

Section 6: Fruit, Vines and Hops

Checklist for Decision Making

Decisions for fertiliser use must be made separately for every field. Where more than one crop is grown in a field, these areas must be considered individually.

1. Confirm the crop to be grown and the intended market. Identify any crop quality requirements for this market.

2. Identify the dominant soil type in the cropped area (see page 86 and Appendix 1).

3. Carry out soil analysis for pH, P, K and Mg before planting and every 3-5 years (see page 35). Target values to maintain when growing fruit, hops or vines are:

   Soil pH 6.0-6.5 (6.5-6.8 before planting),

   Soil P Index 2, K Index 2, Mg Index 2

   Cider apples respond to soil K Index 3 and Mg Index 3.

4. For new plantings and young plantations, identify the SNS Index of the field either by the Measurement Method or by the Field Assessment Method (see pages 95 and 86). SMN analysis is recommended where nitrogen residues are expected to be large (e.g. following previous organic manure use or crops receiving high rates of nitrogen fertiliser).

5. Calculate the nutrients that will be available for crop uptake from organic manures and green waste composts that have been applied since harvest of the previous crop, or which will be applied to the crop being grown (Section 2). Deduct these nutrients from the recommended rates given in the tables.

6. Use regular tissue and fruit mineral analysis for mineral elements to help make fertiliser decisions (see pages 166 and 168). Soil levels are not always reflected in the nutrient concentrations in leaf and fruit.

7. Decide on the strategy for phosphate, potash and magnesium use. This will be either building up, maintaining or running down the soil Index levels (see pages 38-41). Allow for any surplus or deficit of phosphate, potash or magnesium applied to previous crops.

8. Using the tables, decide on the required rate of each nutrient then find the best match for these applications using available fertilisers. Decide on the optimum times for fertiliser application (see page 102).

9. Check that the fertiliser spreader or sprayer is in good working order and has been recently calibrated (see page 49).

10. Keep an accurate record of the fertilisers and organic manures applied.

Section 6: Fruit, Vines and Hops

Fertiliser use for fruit, vines and hops

Soil sampling for pH, P, K and Mg

The results of a soil analysis will only be meaningful if the sample is taken in the field in the correct way and analysed in the laboratory using recognised analytical methods. Recommendations in this *Manual* which are based on soil analytical data are applicable only if the specified methods of sampling and analysis have been followed. In general, laboratory analysis is accurate and the main source of error is when taking the soil sample in the field. Soil analysis results from badly taken samples may be misleading and can result in expensive mistakes if wrong fertiliser decisions are made.

The procedure for sampling arable, vegetable, fruit and grassland is given in Appendix 3. The following modifications should be adopted when sampling fruit, vines or hops.

Sampling before planting

Fields intended for planting should be sampled at 0-15 cm and 15-30 cm soil depths. This is particularly important on land previously in fruit, vines, hops or grass where a depth gradient in nutrient content and acidity will probably have developed. The 15-30 cm sampling is not essential on land that has been ploughed regularly to a depth of 25 cm or more. An even depth of sampling is important, and both the top and bottom portions of the soil core should be included in the sample.

Sampling should be carried out before ploughing so that if lime and fertiliser is needed, it can be applied and then ploughed down. In old herbicide strip orchards, separate samples should be taken from the grass alley and the strip, especially where previous lime and fertiliser applications have been applied to the strip only.

When sampling fields where there is a risk of soil acidity, each core should be tested for pH in the field. Soil analysis of a bulked sample will not necessarily identify acid patches within the field.

Sampling established crops

For all established crops, representative samples should be taken every 3-5 years from the 0-15 cm depth (Appendix 3) but with the following qualifications.

- Orchards in overall grass management or very weedy orchards should be sampled within the spread of the tree branches.
- In orchards with herbicide strip management, sampling should be restricted to the strip, excluding the grass area.
- Samples from soft fruit plantations and hops should be taken at random from within the area of rooting.

Section 6: Fruit, Vines and Hops

Soil acidity and liming

Most fruit crops are tolerant of slight acidity and grow best at around pH 6.0 to 6.5. Soil pH levels below about 5.5 can give rise to manganese toxicity, causing 'measley' bark in apples and purple veining in some strawberry varieties. Blackcurrants are more susceptible to soil acidity and a pH of at least 6.5 should be maintained.

Blueberries are an exception to other fruit as they require a soil pH 4.5-5.5.

Mature hops can tolerate a considerable degree of soil acidity but some varieties may suffer from manganese toxicity if the soil becomes too acid. Young hop plants are more sensitive to acidity.

It is important that soils used for fruit, vines and hops are not over limed as this may lead to micronutrient deficiencies such as iron and manganese.

Liming before planting (see also Section 1 and Appendix 7)

Any lime required should be applied and incorporated before planting. Because acidity problems occur in patches and acidity can develop rapidly when herbicides are used, the whole plough layer should be limed to maintain a pH value of 6.5 in the early years of fruit or hops. Because it will be impossible to correct any acidity at depth by later lime incorporation, the quantity of lime applied before planting should be calculated so that it will correct the pH of the top 40 cm of soil.

Where lime is needed to correct acidity in the subsoil, this lime should be ploughed down. Where sampling has only been carried out to 15 cm depth, the lime requirement using this pH result should be doubled. If the total lime requirement is more than 7.5 t/ha, half should be deeply cultivated into the soil and ploughed down, and the remainder applied and worked in after ploughing. If less than 7.5 t/ha of lime is needed, the whole requirement should be applied after ploughing and cultivated in.

Where there is significant variability of soil pH, lime should be applied at different rates in different areas so that the whole field reaches the same pH. If testing with coloured indicator or a pH meter shows that the soil pH is less than 5.0 below plough depth, seek further advice before liming or planting.

Liming established crops

Under herbicide strip management, the strip will generally become acid more quickly than the grass alley and may require more frequent liming than the alley. The correction of acidity in undisturbed soil is slow, so it is important to check soil pH regularly and apply lime when necessary before the soil becomes too acid and a severe problem builds up.

Liming materials

Acid soils deficient in magnesium may be limed with magnesian limestone, particularly before planting. One tonne of magnesian limestone contains at least 150 kg MgO. However, over-application of magnesian limestone can reduce the availability of soil potash. Where soil

Section 6: Fruit, Vines and Hops

magnesium levels are satisfactory, ground chalk or limestone should be used. The use of coarse grades of limestone or chalk should be avoided.

See Section 1 and Appendix 7 for more information on liming rates and materials.

Magnesium

When the soil also needs liming, magnesium can often be supplied cost-effectively by using magnesian limestone. When liming is not required a magnesium fertiliser should be used.

Where magnesium deficiency has been diagnosed, foliar sprays of agricultural magnesium sulphate (Epsom salts) or other proprietary materials are likely to give a more rapid effect than a soil application of a magnesium fertiliser.

Micronutrients (trace elements)

Micronutrient deficiencies may occur in fruit and vine crops and in hops in some areas, especially where the soil pH is over 7.0. These deficiencies can often be identified by visual symptoms but the diagnosis should be checked by leaf analysis. Iron deficiency cannot reliably be confirmed by leaf analysis.

- Boron (B): Boron deficiency in fruit crops is uncommon, but can occur in hot dry summers, with pears being most susceptible. Where confirmed the deficiency can be corrected by foliar application of boron.
- Copper (Cu): Copper deficiency in pears has been diagnosed on occasions particularly in orchards on sandy soils. It can be corrected by applying a foliar spray of copper.
- Iron (Fe): Iron deficiency occurs commonly in fruit crops grown on shallow calcareous soils. Either soil or foliar application of a suitable iron chelate can be used for treatment.
- Manganese (Mn): Manganese deficiency can occur in fruit crops grown on calcareous soils or soils with a high pH. It is best controlled by foliar application of manganese.
- Zinc (Zn): Zinc deficiency has very occasionally been found to reduce growth and cropping of apple trees on sandy soils. This deficiency can be corrected by foliar application of zinc but applying excessive amounts during blossom or cell division may decrease the number of fruitlets.

Fruit, Vines and Hops – Before Planting

Nitrogen is not required before planting fruit crops, but can be beneficial before planting potted hop cuttings.

Where soil analysis before planting shows soil acidity or low soil P, K or Mg Indices, it is important to correct these shortages by thorough incorporation of appropriate amounts of lime and fertilisers. After planting the downward movement of all nutrients from the soil surface is slow, except for nitrogen. This applies particularly for phosphate and to a lesser extent for potash and magnesium. In organic production systems soil fertility must be built up prior to planting.

Section 6: Fruit, Vines and Hops

Where previously ploughed land has been sampled to 15 cm depth only, the recommended amounts of phosphate, potash and magnesium should be thoroughly incorporated in the autumn before planting. Before planting top fruit, vines or hops, if soil analysis shows the field to be at P, K or Mg Index 0 or 1, the appropriate nutrient amounts should be ploughed down and then the same amount applied again and thoroughly incorporated before planting. If the plough depth is less than 20 cm, the amount ploughed down should be halved.

Where samples have been taken from 0-15 cm and 15-30 cm depths, the recommended rate based on the 15-30 cm sample should be ploughed down before top fruit, vines or hops are planted if the soil P, K or Mg Index is Index 0 or 1. After ploughing, the amount based on the 0-15 cm sample should be applied and thoroughly incorporated. If plough depth is less than 20 cm the amount ploughed down should be halved.

Where it is not possible to plough fertiliser down, the application should be limited to the amount recommended for one sampling depth only. Composted green waste and green manure crops can be incorporated to increase soil organic matter content.

In organic production, derogation is possible for application of supplementary nutrients if soil nutrient Index is low.

The recommendations in this table are based on samples taken from a 15 cm depth of soil

	SNS, P, K, or Mg Index					
	0	1	2	3	4	>4
P mg/l (Olsen's)	0-9	10-15	16-25	36-45	46-70	>71
K mg/l	0-60	61-120	121-240	241-400	401-600	>601
Mg mg/l	0-25	26-50	51-100	101-175	176-250	>251
	kg/ha					
Fruit and vines						
Nitrogen (N)	0	0	0	0	0	0
Phosphate (P_2O_5)	200	100	50	50	0	0
Potash (K_2O)	200	100	50	0	0	0
Magnesium (MgO)	165	125	85	0	0	0
Hops						
Nitrogen (N)	0	0	0	0	0	0
Phosphate (P_2O_5)	250	175	125	100	50	0
Potash (K_2O)	300	250	200	150	100	0
Magnesium (MgO)	250	165	85	0	0	0

- Only use materials containing a large proportion of water-soluble phosphate.
- Potted hop plants benefit from 70-80 kg N/ha applied in the spring before planting.
- Apply this potash in the autumn and thoroughly incorporate it into the soil to avoid root scorch of the newly planted crop.

Don't forget to deduct nutrients applied as organic manures (see Section 2)

Section 6: Fruit, Vines and Hops

Top Fruit, Established Orchards

The nitrogen recommendations are based on the soil management system and soil type. The recommendations are intended as a guide and should be varied according to variety, rootstock, vigour, leaf or fruit analysis and appearance of foliage. Nitrogen dressings can be split across the growing season. The largest demand for nitrogen is between blossom and late July which corresponds with the rapid shoot growth phase and nitrogen applications should reflect this. No application should be made during and after leaf drop.

The results of leaf and fruit analysis are particularly important. The width of the herbicide strip and the effectiveness of the herbicide programme and use of mulches (e.g. straw) can also influence nitrogen requirements. Straw and composted green waste mulches release potash which can antagonise calcium uptake, which in extreme cases can cause physiological fruit disorders where soil calcium availability is low. Guidance on the use of leaf analysis to modify nitrogen and other minerals recommendations is given on pages 166-168.

Applying excess nitrogen encourages vegetative growth with large, dark green leaves. This may adversely affect fruit quality, especially taste, firmness and storage quality. Increasing nitrogen reduces the amount of red colour and intensifies the green colour of apples. This effect is detrimental to crop appearance and value in red coloured varieties, but can be beneficial in culinary varieties such as Bramley. Excess nitrogen can also reduce the storage life of fruit. However, autumn foliar application of nitrogen can improve blossom quality in the following spring.

When nitrogen is deficient, the leaves of fruit crops tend to be small and pale green, the bark of fruit trees may be reddish in colour and shoot growth restricted. Yields are reduced due to the decrease in the number and size of fruit, which may also be highly coloured.

In grass alley herbicide strip orchards, the tree roots are largely confined to the strip and fertiliser should be applied to the herbicide strip only. The nutrient recommendation given in the table are for the complete orchard area and should be reduced pro-rata where nitrogen is applied to the bare soil area only.

Where soil pH is high, consideration should be given to using ammonium sulphate which will help lower the soil pH. Calcium nitrate will have little effect on the soil pH but the calcium applied may improve the storage quality of apples.

Fertigation of young trees

The addition of nutrients to the irrigation water (fertigation) can improve the growth and early cropping of young apple trees planted on sites previously cropped with apples, and may help overcome replant problems. A benefit is more likely where the soil organic matter level and nitrogen reserves have been depleted by long-term use of herbicides intended to maintain a bare soil surface. The rate of nitrogen addition should be about 10 g N/tree in the first growing year, increasing to 15–20 g N/tree in the second and third years. Fertigation will allow fertiliser rates to be reduced by up to 50% of that used for broadcast applications in orchards older than three years. It can also help correct nutrient deficiencies such as phosphate because nutrients in solution are more rapidly moved down the soil profile. Again leaf analysis should be used regularly to provide feedback on adjusting nutrition to appropriate levels.

Care should be taken to ensure soils are not completely wetted to minimise the risk of nitrate leaching.

Section 6: Fruit, Vines and Hops

Biennial fertiliser application

For established crops, the timing of phosphate, potash and magnesium application is not critical.

Established Top Fruit – nitrogen

Crop	Grass/herbicide strip [a]	Overall grass
	kg/ha	
Dessert apples [b]		
Light sand and shallow soils	80	120
Deep silty soils	30	70
Clays	40	80
Other mineral soils	60	100
Culinary and Cider apples		
Light sand and shallow soils	110	150
Deep silty soils	60	100
Clays	70	110
Other mineral soils	90	130
Pears, Cherries and Plums		
Light sand and shallow soils	140	180
Deep silty soils	90	130
Clays	100	140
Other mineral soils	120	160

a. The recommended rates are for the complete orchard area and should be reduced pro-rata where nitrogen is applied to the herbicide strip area only.

b. Larger nitrogen rates may be needed on varieties with regular heavy cropping potential (i.e. >40 t/ha).

Refer to pages 166-168 for guidelines on modifying nitrogen rate according to leaf analysis.

Section 6: Fruit, Vines and Hops

Established top fruit – phosphate, potash and magnesium

	P, K or Mg Index				
	0	1	2	3	4 and over
P mg/l (Olsen's)	0-9	10-15	16-25	26-45	>46
K mg/l	0-60	61-120	121-240	241-400	>401
Mg mg/l	0-25	26-50	51-100	101-175	>176
	kg/ha				
All top fruit, annually					
Phosphate (P_2O_5)	80	40	20	20	0
Potash (K_2O)	220	150	80	0	0
Magnesium (MgO)	100	65	50	0	0

a. Pears require approximately an additional 70kg K_2O/ha up to Index 3, but no addition at Index 4, cider apples also respond to larger applications rates of potash.

The recommended rates are for the complete orchard area. In grass/herbicide strip orchards, the recommended rates should be reduced pro-rata where fertiliser is applied to the herbicide strip area only.

For apples, soil K Index should not be built up above 2 because excessively large potash applications can adversely affect storage quality.

To avoid inducing magnesium deficiency, the soil K: Mg ratio (based on soil mg/litre K and Mg) should be no greater than 3:1. For example, if soil K is 240 mg/litre soil Mg should not exceed 80 mg/litre.

Where the yields of apples and pears are regularly above 40 t/ha, maintenance applications of potash may need to be increased by 20 kg K_2O/ha for every additional 10 t/ha in yield.

When applying phosphate fertilisers, only use those that contain a large proportion of water-soluble phosphate.

Don't forget to deduct nutrients applied as organic manures (see Section 2)

Section 6: Fruit, Vines and Hops

Soft Fruit and Vines – Established Plantations

For bush and cane fruits, nitrogen rates may need to be modified depending on the amount of annual growth required for a particular production system. When nitrogen is deficient, leaves tend to be small and pale green. Nitrogen should not be applied to raspberries after the end of July to avoid excessive growth of soft cane. However, when nitrogen is being applied at lower rates through fertigation, applications may continue up until the end of August. For blackcurrants, fertiliser should be applied to bare soil only or at rates increased to compensate for the grass.

For crops which are establishing prior to reaching full crop potential, smaller rates of nitrogen are usually adequate. The rate should be adjusted according to the amount of growth required and the results of leaf nitrogen analysis.

For established crops, the timing of phosphate, potash and magnesium applications is not critical.

Soft fruit and vines – nitrogen

	kg/ha
Blackcurrants	
Light sand and shallow soils	160
Deep silty soils	110
Clays	120
Other mineral soils	140
Redcurrants, Gooseberries, Raspberries, Loganberries, Tayberries, Blackberries[a]	
Light sand and shallow soils	120
Deep silty soils	70
Clays	80
Other mineral soils	100
Vines[b]	
Light sand and shallow soils	60
Deep silty soils	0
Clays	20
Other mineral soils	40

a. With continuing change in varieties, adjust nitrogen rates depending on plant vigour. b. Excessive growth of vines will cause wood to ripen slowly and a yield reduction in the following crop. Reduce nitrogen rates where growth is excessive.

Section 6: Fruit, Vines and Hops

Soft fruit and vines – phosphate and potash

	P, K or Mg Index				
	0	1	2	3	>4
P mg/l (Olsen's)	0-9	10-15	16-25	26-45	>46
K mg/l	0-60	61-120	121-240	241-400	>401
Mg mg/l	0-25	26-50	51-100	101-175	>176
	kg/ha				
Blackcurrants, Redcurrants, Gooseberries, Raspberries, Loganberries, Tayberries					
Phosphate (P_2O_5)	110	70	40	40	0
Potash (K_2O)	250[a]	180[a]	120	60	0
Blackberries, Vines					
Phosphate (P_2O_5)	110	70	40	40	0
Potash (K_2O)	220	150	80	0	0
All crops					
Magnesium (MgO)	100	65	50	0	0

a. Sulphate of potash should be used for raspberries, redcurrants and gooseberries where more than 120 kg K_2O/ha is applied.

For phosphate fertilisers, use only those that contain a large proportion of water-soluble phosphate.

To avoid inducing magnesium deficiency, the soil K: Mg ratio (based on soil mg/litre K and Mg) should be no greater than 3:1.

Don't forget to deduct nutrients applied as organic manures (see Section 2)

Section 6: Fruit, Vines and Hops

Strawberries – nitrogen

	SNS Index					
	0	1	2	3	4	>5
	kg/ha					
Strawberries – main season						
Light sand and shallow soils	60	50	40	30	20	0
Deep silty soils	0	0	0	0	0	0
Other mineral soils	40	40	30	20	0	0
Strawberries – Everbearers						
Light sand and shallow soils	80	70	60	40	20	0
Deep silty soils	40	30	30	20	0	0
Other mineral soils	60	50	40	20	0	0

With continued change in varieties, adjust nitrogen rates depending on plant vigour and the results of leaf analysis.

Strawberries – phosphate, potash and magnesium

	P, K or Mg Index				
	0	1	2	3	4 and over
P mg/l (Olsen's)	0-9	10-15	16-25	26-45	>46
K mg/l	0-60	61-120	121-240	241-400	>401
Mg mg/l	0-25	26-50	51-100	101-175	>176
	kg/ha				
Phosphate (P_2O_5)	110	70	40	40	0
Potash (K_2O)	220	150	80	0	0
Magnesium (MgO)	100	65	50	0	0

Only use materials containing a large proportion of water-soluble phosphate.

To avoid induced magnesium deficiency, the soil K: Mg ratio (based on soil mg/litre K and Mg) should be no greater than 3:1.

Section 6: Fruit, Vines and Hops

Fertigation

Where strawberries or raspberries are grown under a polythene mulch with sub-irrigation, nutrients can be applied in the irrigation system (fertigation). On soils which encourage vigorous growth, it may be beneficial to reduce nitrogen rates when applied by fertigation. Where growth is not excessive, the nitrogen rates for the whole season should be the same as those recommended for soil applications, but with less being applied during the fruiting period.

At P and K Index 2 or above, maintenance rates of phosphate and potash can be applied by fertigation. However, where the soil P, K or Mg Index is 0 or 1, the recommended amounts of phosphate and potash should be cultivated into the planting bed before the soil is mulched.

Irrigation water may also contain nutrients, particularly calcium, and care should be taken when mixing with fertiliser as insoluble compounds may form which can block emitters.

Don't forget to deduct nutrients applied as organic manures (see Section 2)

Substrate strawberry production

When strawberries are grown in an inert substrate, a complete nutrient solution is required. Normally a conductivity of 1.4 mS/cm is maintained during growth and production for main crop 'June bearers' and the value should not exceed 2.0 mS/cm. High salinity can cause marginal necrosis and stimulate leaf and flower tip burn. During vegetative growth the soil K: Ca ratio (based on mg/litre K and Ca) should be maintained at 0.65, and at 0.8 during flowering and fruiting to improve fruit taste and firmness. Plants grown on substrates are very sensitive to excessive concentrations of zinc, boron and sodium in the nutrient solution. Deficiency of iron and manganese can occur at high (alkaline) pH levels in the substrate.

The nutrients will need to be adjusted depending on whether peat or coir substrates are used. Coir is usually supplied unfertilised and therefore needs wetting up before planting with a feed solution for 2 – 3 days. It needs more calcium, magnesium and sulphur, but less boron and potassium when used fresh. Owing to its inherently high pH, coir needs a lower solution pH (5.3 – 5.8) than for peat pH (5.6 – 6.0). Furthermore, feed recipes also depend on the chemical composition of the irrigation water and should be modified during the growing season according to the results of substrate, leaf tissue and drainage solutions analyses.

Section 6: Fruit, Vines and Hops

Guidelines for nutrient solution for strawberry production on substrate

	Nutrient solution optimum mg/litre in dilute feed		
Nitrogen (NO_3)	110-140	Iron (Fe)	1.1-1.7
Nitrogen (NH_4)	7-14	Zinc (Zn)	0.46-0.65
Phosphorus (P)	46	Boron (B)	0.11-0.17
Potassium (K)	140-250	Manganese (Mn)	0.55-1.11
Magnesium (Mg)	30-40	Copper (Cu)	0.03
Calcium (Ca)	140-180	Molybdenum (Mo)	0.05
Sulphate (SO_4)	50-100		

Leaf Analysis for Top and Soft Fruit

Leaf analysis is an essential technique for general monitoring of nutrient status and the diagnosis of nutritional disorders. Separate samples should be taken in a similar manner from good and poor areas of growth so that the results can be compared.

In addition, knowledge of leaf nutrient concentrations has proved useful for assessing the nutritional status of crops. Satisfactory ranges for optimal growth and cropping are given in the table below. Where analysis results are to be compared to these standards, it is essential that a representative sample is taken, in the correct way and at the correct time.

Because there are seasonal and other factors which influence leaf nutrient concentrations, leaf analysis must be interpreted carefully. Leaf nutrient levels can also vary between varieties. Where there is sufficient information, the standard ranges take account of differences between varieties.

Leaf analysis can be used to provide a more complete guide to the adequacy of the orchard fertiliser programme than can be obtained from soil analysis alone. Where leaf nutrient levels are below the satisfactory range, an increase in fertiliser use can be considered. However, before making a change, the cause of the problem should be further investigated to ensure that other factors such as soil compaction or disease are not involved.

Where the leaf nutrient level is consistently above the satisfactory range for several years, there is justification for a reduction in fertiliser use. In particular, high levels of nitrogen and potash can have adverse effects on apple storage quality and application rates can often be reduced to advantage.

A high manganese level indicates a need to check soil pH, as it is often associated with increased soil acidity, but can also result from use of foliar feeds or fungicides containing manganese.

Section 6: Fruit, Vines and Hops

Leaf analysis – satisfactory nutrient ranges of major nutrients expressed as elements

Crop	Leaf sampling position[a]	Nitrogen (N)	Phosphorus (P)	Potassium (K)	Magnesium (Mg)	Sulphur (S)
		% in dry matter				
Apple						
Cox[b]	1	2.6-2.8	0.20-0.25	1.2-1.6	0.20-0.25	0.20-0.40
Bramley	1	2.4-2.8	0.18-0.23	1.2-1.6	0.20-0.30	
Cherries	1	2.4-2.8	0.20-0.25	1.5-2.0	0.20-0.25	0.13-0.84
Pears						
Comice	1	1.8-2.1	0.15-0.20	1.2-1.6	0.20-0.25	0.17-0.26
Conference	1	2.1-2.6	0.15-0.20	1.2-1.6[c]	0.20-0.25	
Plums	1	2.0-2.6	0.15-0.20	1.5-2.0	0.20-0.25	0.20-0.70
Blackcurrants	2	2.8-3.0	0.25-0.35	1.5-2.0	0.15-0.25	
Raspberries	3	2.4-2.8	0.20-0.25	1.5-2.0	0.30-0.35	
Strawberries	4	2.6-3.0	0.25-0.30	1.5-2.0	0.15-0.20	0.10-0.20
Vines	5	2.0-3.0	0.25-0.30	1.2-1.6	0.20-0.30	
Blueberries	6	1.8-2.0	0.08-0.4	0.4-0.7	0.13-0.25	0.12-0.2

a. Leaf sampling position
   1 Mid third extension growth, sampled mid-late August.
   2 Fully expanded leaves extension growth, sampled prior to harvest.
   3 Fully expanded leaves non-fruiting canes, sampled at fruit ripening.
   4 Lamina of recently matured leaves, sampled at fruit ripening.
   5 Leaf opposite basal fruit cluster, sampled at full bloom.
   6 Fully expanded leaves between late July and mid August

b. For Gala and Braeburn follow Cox, however, typical average for Gala is 2.3% N and P content is less than in Cox.

c. Yield benefits are achieved at 1.6 % K

Section 6: Fruit, Vines and Hops

Leaf analysis – satisfactory nutrient ranges of micronutrients

Micronutrient	Deficiency	Optimum	High
	mg/kg in dry matter		
Manganese (Mn)	20	30-100	100[a]
Boron (B)[b]	15	20-40	40[c]
Zinc (Zn)	10	15-30	50
Copper (Cu)	5	7-15	15
Iron (Fe)	<45	45-250	

a. Manganese concentrations above 100 mg Mn/kg indicate that the soil is becoming acid. Check the soil pH.

b. Fruit analysis is the most reliable diagnostic technique for boron deficiency. Optimum levels are 1.5 to 4.5 mg B/kg fresh weight. Below 1.5 mg B/kg indicates deficiency.

c. Excess boron levels can promote premature ripening and senescence in fruit.

Apple Fruit Analysis

Analysis of fruit sampled within three weeks of picking is a useful indicator of the risk of some physiological disorders in stored apples. Results can also be used to rank orchards for potential storage quality. During the period between sampling for analysis and harvesting the fruit, the concentration of calcium falls mainly due to dilution as fruit size increases. Analysing fruit too far in advance of harvest may over-estimate the storage potential.

Fruit samples should be taken as near to harvest as possible but within 2 weeks of picking. In each orchard randomly select 30 trees of the same age and variety. Take one apple at random from each tree, alternating from side to side and at different heights but ignore abnormally large or small fruits. Try to make the sample representative of the side of trees where most fruit is growing.

If fruit analysis produces consistently high or low concentrations of a particular nutrient over two to three years, modification of fertiliser application should be considered. The most likely change will be a reduction in nitrogen or potash use. Fruit analysis may also show deficiencies of calcium or phosphorus which can reduce fruit storage quality. These deficiencies can be corrected by foliar sprays of calcium and phosphorus or by post-harvest calcium treatments.

Section 6: Fruit, Vines and Hops

Apple Fruit analysis – satisfactory nutrient concentrations for storage (sampled at harvest)

Crop	Nitrogen (N)	Phosphorus (P)	Potassium (K)	Magnesium (Mg)	Calcium (Ca)
	mg/100g fresh weight				
Cox	50-70	11.0 minimum	130-150	5.0	4.5 minimum[a] 5.0 minimum[b]
Bramley	60 minimum	9.0 maximum	105-115	5.0	4.5 minimum[a] 5.0 minimum[b]

a. For controlled atmosphere storage (Cox in 2% oxygen until late February or 1.2% oxygen until late March; Bramley in 8–10% carbon dioxide until June or 5.0% carbon dioxide plus 1.0% oxygen until July).

b. For storage in air at recommended temperature (Cox until mid October; Bramley until November).

Satisfactory nutrient concentrations have not been established for other varieties, but the calcium requirements are likely to be similar to those given in the table. The standards given for calcium and potassium concentrations in Cox also apply to Egremont Russet. Average values across many orchards for Gala and Braeburn apples are given below.

Average nutrient concentrations in Gala and Braeburn apples

Crop	Nitrogen (N)	Phosphorus (P)	Potassium (K)	Magnesium (Mg)	Calcium (Ca)
	mg/100g fresh weight				
Gala 2 wk < harvest	45	10.2	106	5.3	9.6
Gala harvest	42	9.3	122	5.0	7.4
Braeburn 2 wk < harvest	52	11.9	102	5.1	6.3

Notes

- Gala is naturally low in P compared to Cox, but this does not affect storage potential
- Gala is naturally high in Ca and does not suffer from Ca-dependent storage disorders
- Large concentrations of K in Gala fruit may increase the risk of breakdown
- Gala fruits higher in Ca and lower in K may be firmer ex-store
- Braeburn has a similar composition to Cox except that K concentration appears lower
- Suggest using Cox calcium threshold concentrations to judge storage potential of Braeburn apples

Section 6: Fruit, Vines and Hops

Nitrogen (N): As the nitrogen content increases, fruit becomes more susceptible to rotting, loss of firmness, poor skin finish and a lack of red colour. Above 80 mg N/100g the risk of disorders in Cox is high. In Bramley, a large nitrogen concentration leads to more green colour, but risk of lower firmness.

Phosphorus (P): At phosphorus contents below 11 mg P/100g in Cox, there is an increased risk of fruit losing firmness and developing breakdown, particularly if calcium is also low. In Bramley, the phosphorus threshold for breakdown is lower at 9 mg P/100g.

Potassium (K): A high potassium content will increase the risk of bitter pit, *Gloeosporium* rotting and core flush. The risk of bitter pit is also greater if the calcium level is low in relation to potassium. Generally, fruit flavour and acidity increases with increasing levels of potassium. Thus, if calcium levels are adequate (over 5.5 mg Ca/100g), large concentrations of potassium may be advantageous in terms of fruit quality.

Magnesium (Mg): A high magnesium content will increase the risk of bitter pit, especially when calcium levels are marginal.

Calcium (Ca): Calcium levels of 5.0 mg Ca/100g and above are necessary to maintain high quality throughout long-term storage. However, the storage potential will be modified depending on the content of other elements especially potassium and phosphorus. Disorders associated with low calcium (bitter pit in particular) are more likely to occur in apples of a given calcium content which are stored in air rather than controlled atmosphere. Thus, fruit which meets the standards for nitrogen, phosphorus, potassium and magnesium and has a calcium concentration in the range of 4 to 5 mg Ca/100g should not be rejected for mid-term controlled atmosphere storage as it is unlikely to develop commercially significant levels of bitter pit or breakdown. This much lower risk is reflected by the dual standards for calcium given in the table.

The risk of bitter pit and susceptibility to *Gloeosporium* will depend on the ratio of K: Ca. If the K: Ca ratio is over 30:1 in air stored Cox or Bramley stored in controlled atmosphere, or over 25:1 in air-stored Bramley, commercially important losses due to bitter pit are likely. Where calcium contents are marginal (3.5–4.0 mg Ca/100g) and fruit phosphorus is also less than 9.0 mg P/100g, both Cox and Bramley are more susceptible to breakdown. In such cases, Cox should be marketed early. Bramley should be stored at a higher temperature (average 4.5°C) and sold earlier than fruit with optimum levels of calcium and phosphorus.

Gloeosporium risk is dependent on the level of inoculum in the orchard and is influenced by skin finish, fungicide programme and rainfall prior to harvest. Fruit analysis is a measure of fruit susceptibility to *Gloeosporium* and not necessarily the eventual amount of rotting.

The incidence of senescent breakdown is greatest in late picked fruit for any given content of phosphorus and calcium.

Susceptibility to low temperature breakdown in Cox stored at 3.0°C (air) and 3.5°C (controlled atmosphere), although less common in Bramley at 4.0°C, is also partly due to low calcium and phosphorus contents.

Section 6: Fruit, Vines and Hops

Hops

(See pages 157-158 for pre-planting fertiliser recommendations.)

Fertiliser is not required in the establishment year provided appropriate pre-planting fertilisers have been applied.

Established Hops (second and subsequent years after establishment) – Nitrogen

	kg/ha
Deep silty soils	180
Clay soils	200
Other mineral soils	220

The recommended rates are for maximum yield situations and should be applied annually. Nitrogen can reduce the alpha-acid content of hop cones, though it may produce more alpha-acid per hectare because the crop yield is greater. Where progressive *Verticillium* wilt is present, high nitrogen rates will make hops more susceptible to this disease so reduce the recommended amount to 125–165 kg N/ha where there is a risk of wilt.

Nitrogen should be split into two or three applications, the first dressing being given in late March or April, the second during May and the third in late June or early July. There is some evidence that late hop varieties especially, respond to a three timing split, with the last application no later than early July. The total rate should be adjusted according to variety, irrigation and soil type.

Where trickle irrigation is used there is benefit in using fertigation to apply nutrients.

Where large and frequent applications of organic manures have been used in previous years, reduce the nitrogen recommendation by 70 kg N/ha. FYM is best applied in the early summer to help minimise aggravating the effect of any wilt present, with alleyways treated in alternate years. Where organic manures have been applied in the previous 12 months the nitrogen rate should be reduced according to the information in Section 2.

Where soil nitrogen residues are potentially large (e.g. where organic manures have been used regularly) the use of soil analysis for SMN can be useful. See pages 95-96 for further details and Appendix 2 for sampling guidelines.

Section 6: Fruit, Vines and Hops

Established hops – phosphate, potash and magnesium

	P, K or Mg Index					
	0	1	2	3	4	5 and over
P mg/l (Olsen's)	0-9	10-15	16-25	26-45	46-70	>71
K mg/l	0-60	61-120	121-240	241-400	401-600	>601
Mg mg/l	0-25	26-50	51-100	101-175	176-250	>250
	kg/ha					
Phosphate (P_2O_5)	250	200	150	100	50	0
Potash (K_2O)	425	350	275	200	100	0
Magnesium (MgO)	150	100	50	0	0	0

Hops require the maintenance of large soil nutrient reserves, P Index 4, K Index 3 and Mg Index 2. Potash is important and care must be taken to ensure that the recommended rates are applied annually.

Phosphate fertilisers should contain a large proportion of water-soluble phosphate.
To avoid induced magnesium deficiency, the soil K: Mg ratio (based on soil mg/litre K and Mg) should be no greater than 3:1.

Farmyard manure

Farmyard manure has been traditionally used on hops. As well as supplying nutrients it helps to improve the structure of cultivated soils. Now that few soils growing hops are cultivated there is less need for regular applications of bulky organic manures. Farmyard manure is recommended where the soil continues to be cultivated and where land is being prepared for planting.

Extreme caution should be exercised in the use of farmyard manure or slurry where *Verticillium* wilt is known or suspected to be present. Heavy applications of manure, in addition to supplying excess nitrogen, can reduce the soil temperature during the critical spring period. Low soil temperatures in the spring are known to make hops more susceptible to the disease.

Don't forget to deduct nutrients applied as organic manures see (Section 2)

Section 7: Biomass crops

In the UK two crops are being grown commercially specifically for biomass for use as a source of energy – willow (*Salix* spp) grown as short-rotation coppice (SRC) and Miscanthus (elephant grass, normally *Miscanthus x giganteous*). Other possible biomass crops such as switchgrass and poplar have only been grown in the UK under experimental conditions.

In addition to these dedicated biomass crops, some normal agricultural crops are also grown for energy purposes: these include wheat grain and sugar beet, both for bioethanol, and oilseed rape for biodiesel. Fertiliser requirement for these crops, if grown for energy use, is covered under the individual crops. Various agricultural by-products may be used for energy generation, such as incineration of cereal straw for electricity generation or anaerobic digestion of animal slurry for biogas generation.

One of the main reasons for growing crops for energy production is to replace some usage of fossil fuels and thus decrease greenhouse gas (GHG) emissions. Thus it is logical to ensure that GHG emissions associated with growing energy crops, directly or indirectly, are kept to a minimum. Nitrogen fertilisers lead to large emissions of GHGs – carbon dioxide (CO_2) and nitrous oxide (N_2O) during manufacture and additional N_2O when applied to soil. N_2O is a particularly powerful GHG, each molecule having the same greenhouse-warming potential as about 300 molecules of CO_2. In total, the manufacture and use of each tonne of nitrogen fertiliser is equivalent to at least 5 tonne CO_2-equivalent. So it is important to consider the environmental effects of nitrogen fertiliser applications when growing crops for energy.

Sources of information

Only a small number of studies on the nutrient requirements of either Miscanthus or SRC willow have been conducted, either under UK conditions or worldwide. With such a small knowledge base it is difficult to be certain of the extent to which the individual results can be generalised, or whether they are site-specific. Consequently the guidance given here is preliminary and based largely on replacing nutrients removed in the harvested crops unless otherwise stated.

Miscanthus

Typical offtake of nutrients in harvested biomass (excluding the first two years after planting when yields are much lower than in later years) are:

Nutrient	Per tonne dry biomass	In a typical crop yielding 14 tonne dry biomass per ha
	kg/t	kg/ha
N	6	84
P_2O_5	1	14
K_2O	8.5[a]	120[a]

a. Potash offtake in Miscanthus is very variable being affected by weather and time of harvesting. These values refer to crops harvested in January; it is now common for crops to be harvested later (e.g. April-May) and the offtake of potash then is generally less because rainfall leaches out potash from the standing crop and returns it to the soil.

Section 7: Biomass crops

There have been no published studies to test crop responses to different applications of phosphate or potash, and only a few with nitrogen. Nitrogen and phosphate offtakes are small compared with may other crops because much nitrogen and phosphate is transferred from stems and leaves to the rhizomes before harvest; these nutrients can be re-used in future years. In addition to nutrients removed in the harvested parts of the crop, an additional amount is required for the growth of rhizomes in the first few years after planting. On the basis of (a) a relatively small number of measured nutrient offtakes by Miscanthus, (b) comparisons with offtakes by other crops on soils at different levels of crop-available phosphate or potash, and (c) experience gained by those growing the crop the following is proposed:

Phosphate

Maintain soil at P Index 1. Check every 3-5 years by soil testing (see Section 1 and Appendices 3 and 4).

Potash

Maintain soil at K Index 1- 2. Check every 3-5 years by soil testing (see Section 1 and Appendices 3 and 4).

Nitrogen

In some cases a biomass yield response to 50-100 kg N/ha has been observed, but rarely if ever to higher rates and sometimes no response for many years (up to 15 years at one UK experiment). For Miscanthus growing on soil previously under arable crops, and with little previous organic manure (i.e. in SNS Index 2 or below), the soil is likely to supply about 40 kg N/ha (depending on soil type and management history). On the basis of current information, annual fertiliser applications in the range of 60-80 kg N/ha (or organic applications estimated to supply this quantity of nitrogen) are likely to provide sufficient nitrogen for maximum production. In soils starting at a higher SNS Index, nitrogen applications are probably not required for some years.

In the first 2-3 years after planting, some nitrogen is required for the growth of rhizomes in addition to that removed in harvested biomass. However, the quantity of nitrogen removed in these years is less than in subsequent years so it is likely that no additional nitrogen for rhizome development is required. Thus, in contrast to some earlier suggestions, it is recommended that very little nitrogen (perhaps none) will generally be required in the first 2 years; nitrogen applications, as inorganic fertiliser or organic manures only should start for the third year's crop. There is some evidence that nitrogen applications applied in the first year after planting are subject to large losses and also encourage weed growth.

Applying nitrogen in late May, just before rapid growth begins, is common practice and this seems appropriate – though there has been no work to test the best time to apply nitrogen.

Sulphur

There is currently no evidence of sulphur applications being required by Miscanthus in the UK.

Section 7: Biomass crops

Willow

Typical offtakes of nutrients in wood harvested from willow SRC after 3 years growth following the previous coppicing are.

Nutrient	Per tonne dry biomass	In a typical crop yielding 30 tonne dry biomass per ha
		kg/ha
N	3	90
P_2O_5	1.8	55
K_2O	2.4	72

For SRC willow there is even less experimental data on nutrient requirements than for Miscanthus. So the provisional recommendations given here are based a combination of nutrient offtakes measured at a very small number of sites and experience of growers. No fertilisers are required during the establishment year; so the recommendations below refer to the three year periods after the initial cutback (in the winter after establishment) and the subsequent three year periods after each harvest. With SRC it is almost impossible to enter the plantation to apply fertilisers or manures in the second or third year after cutback or harvesting unless specialist equipment is available. It is therefore necessary to determine nutrient requirements, and make any appropriate applications, in the first year after a cutback, normally during spring in preparation for the period of rapid growth.

In situations where biomass crops are used for bioremediation of biosolids or bio filtration of effluents, levels of nutrient application in excess of those highlighted above may be justified, recognising the relatively low environmental risks associated with this approach, relative to alternative methods of disposal. In such circumstances, careful monitoring of ground and surface water is required to demonstrate absence of leaching and/or surface runoff of N and P.

Phosphate

Maintain soil at P Index 1. Check every 3 years (i.e. each harvesting cycle) by soil testing (see Section 1 and Appendices 3 and 4).

Potash

Maintain soil at K Index 1. Check every 3 years (i.e. each harvesting cycle) by soil testing (see Section 1 and Appendices 3 and 4).

Section 7: Biomass crops

Nitrogen

As with Miscanthus, a combination of nitrogen mineralised from soil organic matter plus that from atmospheric deposition and rain will supply much of the modest nitrogen requirement of about 30 kg N/ha per year for a typical UK crop yielding 30 t dry biomass during a three year cycle, i.e. 90 kg N/ha total offtake. However, larger offtakes have been observed in some situations and, as a grower gains experience of yields in specific soil types and different environments, nitrogen applications can be adjusted using the values in the table above for nitrogen removed per tonne dry biomass.

Applications of organic materials such as FYM, slurry or sewage sludge are an ideal way of maintaining a suitable supply of nitrogen and other nutrients throughout the three year growth cycle from a single application made in the first year. But amounts applied should be determined on the basis of estimated crop requirement, or within the limits described above for bioremediation or bio filtration systems.

Section 8: Grass

	Page
Differences between 7th and 8th editions of grasslands chapters of (RB209)	178
Check list for decision making	179
Principles of fertilising grassland	181
Protecting the environment	185
Finding the nitrogen recommendation	187
Finding the phosphate, potash and magnesium recommendation	190
Grass establishment – nitrogen	191
Grass establishment – phosphate, potash and magnesium	191
Cutting and Grazing for dairy production – nitrogen	192
Cutting and Grazing for beef production – nitrogen	198
Cutting and Grazing for sheep production – nitrogen	203
Grazing of grass/clover swards – nitrogen	208
Grazed grass – phosphate, potash and magnesium	210
Cutting of grass/clover swards, red clover and lucerne – nitrogen	210
Grass silage – phosphate, potash, magnesium and sulphur	211
Hay – nitrogen	213
Hay – phosphate, potash and magnesium	213

Recommendations for nitrogen (N), phosphate (P_2O_5), potash (K_2O), magnesium (MgO), sulphur (SO_3) and sodium (Na_2O) are given in kilograms per hectare (kg/ha).

The recommendations are based on the nutrient supplies needed to provide the herbage requirement of a grassland system. Allowance is made for the nutrients already contained in the soil. More background on the principles underlying the recommendations is given in Section 1 'Principles of Nutrient Management and Fertiliser Use'.

The method for assessing the reserves of nitrogen in grassland soils is different to the Soil Nitrogen Supply Index (SNS) system used in arable or vegetable systems.

Section 8: Grass

Differences between 7th and 8th editions of grasslands chapters of (RB209)

- The nitrogen recommendations are based on a new analysis of existing national multi-site nitrogen response data for UK grassland.
- The overall approach is no longer totally based on the Nopt for grass growth. The new version is based on the need to supply sufficient home grown forage for particular animal production systems (dairy, beef and sheep) at different levels of intensity of production, stocking rate, and concentrates use.
- This a farm systems based approach.
- The new recommendations enable farmers who may be operating at widely different stocking rates and feeding different levels of concentrates to obtain relevant recommendations for whole season total nitrogen requirement.
- Nitrogen requirements have been calculated using commonly accepted values of daily liveweight gain, forage intake and feed energy conversion to milk, proportions of available land devoted to cutting and grazing and efficiencies with which home-grown forage can be utilised under the two harvesting regimes (cut and grazed grass).
- Whole season nitrogen requirements are provided in groups of tables for each of the main livestock enterprises, dairy, beef and sheep. The first categorisation in each livestock section is by Grass Growth Class, where recommendations are provided for both cut and grazed grass.
- The recommendation tables indicate how it is possible to decrease total nitrogen requirements the better the Grass Growth Class in order to produce the target amount of home grown forage, as nitrogen is used more efficiently. In other words, the better the Grass Growth Class, the greater the efficiency of nitrogen use and a reduced risk of nitrogen losses to the environment. Conversely, the poorer the Grass Growth Class the higher the nitrogen requirement (assuming no other limiting factors, such as shortage of phosphate, potash and sulphur supply). This reflects the need to support a target economic level of animal production rather than maximum yield of forage.
- There is improved guidance on nitrogen contribution from clover, based on photographic recognition of percentage cover in the sward.
- The intervals between the indices for phosphate and potash have been increased to 30 kg/ha (from 25 kg/ha) to bring the rationale in line with arable soils.

Section 8: Grass

Check list for Decision Making

The following checklist provides a convenient framework for making accurate fertiliser decisions.

It is important that individual decisions are made for each field on the farm each year.

Planning grassland management

1. Plan the stocking rate, and cutting and grazing strategy for the farm, and decide if reliance is to be made on clover as a source of nitrogen. This strategy will need to take account of the Grass Growth Class of the farmland (see page 188). Since the amount of grass growth can be very seasonally dependent, this strategy should be regularly reviewed to adjust for variations in growth as the season progresses.

2. For each field, plan the amounts and timings of fertiliser applications that are likely to be needed. Remember that nitrogen applications may need to be adjusted depending on the season and recent weather conditions.

Nitrogen

1. Determine the Soil Nitrogen Supply status of the field (see page 187). Most intensively-managed grassland has a high Soil Nitrogen Supply due to dung and urine returned to the soil during grazing or from application of organic manures.

2. Identify the Grass Growth Class of the field (see page 188).

3. Use the tables and accompanying notes on pages 191-213 to decide on the nitrogen rate and timing for grazing or cutting. The recommendations give the average nitrogen rates needed to grow grass on the farm for the levels of stocking and feeding intensity specified and according to Grass Growth Class. Little or no nitrogen should be used in most grass/clover systems.

4. Adjust the recommendation to take account of the available nitrogen supplied from any organic manures applied for that cut or grazing (see page 182 and Section 2).

5. Review mid-season use of nitrogen depending on actual grass growth and particularly in droughty conditions (see page 189).

6. Do not apply nitrogen after mid August except where autumn growth is required in grass/clover systems.

Lime, phosphate, potash, magnesium and sulphur

1. Carry out soil analysis for pH, P, K and Mg every 3-5 years (see page 35). Target values to maintain in continuous grassland or ley/arable systems are:
   Soil pH 6.0 (5.3 on peat soils) for continuous grass
   Soil pH 6.2 (5.6 on peat soils) where an occasional barley crop is grown
   Soil pH 6.0 (5.3 on peat soils) where an occasional wheat or oat crop is grown
   Soil P Index 2
   Soil K 2-
   Soil Mg Index 2

June 2010

Section 8: Grass

2. Decide on the strategy for phosphate and potash use. This will be building up, maintaining or running down the soil Index (see pages 38-41). Allow for any surplus or deficit of phosphate or potash from previous manure and fertiliser applications.

3. If a reasonably accurate yield estimate can be made, consider calculating the amount of phosphate and potash removed in silage or hay crops over the whole season (see page 211 and Appendix 5). This is the amount of these nutrients that must be replaced in order to maintain the soil Index.

4. Adjust the recommendations to take account of the nutrients in any organic manures applied (see page 182 and Section 2).

5. Assess the need for sulphur fertiliser. Many silage crops, especially second or later cuts, will benefit from using a sulphate containing fertiliser (see page 212).

All nutrients

1. Select a compound or straight fertiliser that matches as closely as possible the nutrient requirements calculated above. It is most important to get the rate and timing of nitrogen correct. The exact rate and timing of other nutrients is usually less critical unless the soil Index is low.

2. Check that the fertiliser spreader or sprayer is in good working order, is serviced annually and is calibrated every spring and whenever fertiliser type changes. Tray-test the spreader annually (see page 49). Check the manure spreader for mechanical condition and calibrate annually.

3. Keep a clear record of the fertilisers and manures applied.

Section 8: Grass

Principles of Fertilising Grassland

The nutrition of grassland is more complex than that of arable crops but the main features and principles are well established. Compared to arable cropping, more nutrient sources must be taken into account and, almost always, manures are applied to part of the grassland area. Most importantly, herbage is produced to meet the needs of livestock on the farm and there is no point in producing more than can be utilised effectively. In many cases, this means that the full productive potential of grazed grassland is not needed.

Most fertiliser policies on grassland farms need to integrate the use of inorganic fertilisers with maximising the value of the readily available nutrients contained in organic manures. Manure is returned to grassland in two ways: by application as FYM or slurry and by deposition of dung and urine by grazing animals. Large proportions of the nutrients in herbage eaten by grazing animals are returned to the sward by deposition. This recycling is not efficient as application rates within dung and urine patches are much larger than grass requirement, significant losses of nitrogen as ammonia occur from urine patches and a small proportion of the sward is affected by dung or urine at any given time. However, over several years, nutrient reserves tend to accumulate in grazed land.

Slurry and FYM are valuable sources of nutrients and their correct use can result in substantial savings in inorganic fertilisers with reduced risk of causing environmental pollution. Section 2 of this Manual gives details of how to make best use of manures and how to estimate the fertiliser value of individual applications.

Nitrogen

Swards that are mainly ryegrass respond strongly to increasing nitrogen supply. Response tends to occur in two stages: firstly, nitrogen is taken up and secondly, dry-matter yield increases. Nitrogen uptake is more rapid than yield increase and is less affected by some adverse conditions such as short day length. It is necessary therefore to distinguish greening of grass (associated with nitrogen uptake) from herbage growth.

Although nitrogen has a very large effect on the growth of grass swards, target dry matter yields will only be obtained if other requirements are met:

- adequate supply of moisture (rainfall or stored soil moisture)
- adequate temperature
- balanced supply of other plant nutrients
- satisfactory soil pH
- satisfactory sward composition

Recommendations in this Section apply where these conditions are met. Where yield is severely restricted by any of these factors nitrogen inputs from fertilisers or manures should be reduced. Summer drought is the most common cause of poor growth either due to seasonally low rainfall or where there is a drought-prone soil type. When summer growth is restricted due to lack of rain, subsequent fertiliser inputs should be reduced or omitted since previous applications will not have been fully used by the crop.

June 2010

Section 8: Grass

In most seasons, growth up to late spring will not be restricted by moisture stress even on the most drought-prone soils. This also coincides with the period of most rapid growth. Therefore, a high proportion of the total nitrogen use in intensive systems should be applied in spring and early summer. This is particularly important on drought-prone soils since summer growth cannot be guaranteed. Nitrogen should not normally be applied after August.

Patterns of nitrogen application must be matched to the ability of the system to utilise the grass grown and to the quality of grass required. Guidance is given as footnotes to the recommendation tables but a high degree of local judgement is also required. Grass is only of value if it can be utilised by livestock. It is not sensible to grow more grass than is needed for the livestock on the farm or if the land is too wet (e.g. in spring) to utilise the grass effectively by cutting or grazing.

In grassland systems, the amount of nitrogen supplied from soil reserves varies considerably. The fertiliser nitrogen values recommended refer to moderate Soil Nitrogen Supply status and adjustments must be made for higher or lower Soil Nitrogen Supply in a particular field. Many intensively managed swards, except newly established leys, will have a high soil nitrogen status which will result in a lower requirement for fertiliser nitrogen. This is because:

- soil nitrogen builds up as a result of nitrogen returned in organic manures or excreta during grazing
- the perennial plant cover limits losses of nitrogen by leaching
- established grassland is not ploughed which limits nitrogen losses

The effect of regular applications of organic manures

Livestock manures are an important source of valuable nutrients and need to be recycled to land. Careful and planned use of manures can result in large savings in purchased fertilisers.

Manures supply both readily available nitrogen and nitrogen in organic form that is released only slowly for crop uptake. The readily available nitrogen content of a manure application will usually be taken up by grass, or lost to the environment, during the first season following application. Section 2 gives information on how to calculate the available nitrogen that is equivalent to fertiliser nitrogen. The remaining organic nitrogen will be added to the soil organic matter pool and will contribute to the supply of soil nitrogen for several seasons (i.e. it will increase the soil nitrogen status). When deciding on fertiliser nitrogen use, it is important to consider both the supply of available nitrogen from recent manure applications and the available nitrogen released from older applications.

In cattle and sheep systems, most of the nitrogen ingested as conserved grass or feed, is excreted. During grazing, this nitrogen is returned to the soil and, in intensive systems, will result in the soil having a moderate or high soil nitrogen status. During housing, excreta are collected as manure. If this manure is spread equally to all of the grass conservation fields that produced the forage, then most of the nitrogen removed from the soil during the growing season will be replaced. Although this nitrogen will not all be immediately available for uptake from an individual manure application, the total available nitrogen contribution from a past history of regular manure applications will be substantial.

Section 8: Grass

Thus, silage fields that receive regular average applications of manure will usually have a moderate or high soil nitrogen status. Fields which are regularly cut for silage **and** receive little or no manures are likely to have a low soil nitrogen status. Adjustments upwards to higher rates of nitrogen may therefore be justified.

The nutrient recommendations in the tables take account of past manure applications because these influence assessment of the soil nitrogen status of a field. The available nitrogen content of manures recently applied for the current season's growth (i.e. previous September applications onwards) must be assessed using the information in Section 2 and deducted from the table recommendation.

Phosphate and potash

The general principles of using phosphate, potash and magnesium fertilisers are described in Section 1. As described above for nitrogen, the recycling of phosphate and potash through conserved grass and back to silage fields in manure applications will commonly result in sufficient nutrients to maintain soil P and K Indices.

Fields which are regularly cut for silage **and** receive little or no manures will have a higher requirement for phosphate and potash. Regular soil analysis every 3-5 years and using the information in Section 2 will ensure that the nutrients in manures are used profitably and with minimal risk of causing pollution.

For grassland the following additional points should be noted:

- Grass/clover swards are more sensitive to phosphate and potash shortages than pure grass swards.
- Grass silage can remove large quantities of potash which must be replaced by fertiliser or organic manure application. Failure to do this can lead to rapid development of potash deficiency and low grass yields.
- It is important to maintain a suitable balance of nutrients in grass, particularly when it is grazed. Nutrient imbalances can aggravate livestock disorders such as a shortage of magnesium (hypomagnesaemia or grass staggers). The risk of hypomagnesaemia can be reduced by applying adequate amounts of magnesium as fertiliser and/or dietary supplements, and by avoiding excessive use of potash fertiliser. Avoid applying potash fertiliser in spring to grazing land except at soil K Index 0.
- Early spring growth can benefit from a small quantity of spring-applied phosphate. All or part of the total phosphate requirement should be applied in early spring.

Recommendations are given as kg/ha of phosphate (P_2O_5) and potash (K_2O). Conversion tables are given in Appendix 8.

Magnesium

Although grass growth is unlikely to respond to magnesium fertiliser, it is important to maintain an adequate level of magnesium in grass herbage to help minimise the risk of livestock disorders such as hypomagnesaemia (grass staggers). Direct treatment of stock may also be needed to avoid this disorder.

Section 8: Grass

At soil Mg Index 0, apply 50 to 100 kg MgO/ha every three or four years. If there is a risk of hypomagnesaemia, larger amounts may be justified to maintain soil Mg Index 2. Where liming is also needed, use of magnesian limestone may be most cost effective.

Magnesium recommendations are given as kg/ha of magnesium oxide (MgO) not as Mg. Conversion tables are given in Appendix 8.

Sulphur

Sulphur deficiency is increasingly common in grassland, especially at second and later cuts in multi-cut silage systems using high rates of nitrogen, but also sometimes at first cut. Deficiency can cause large reductions in yield. From a distance, visible symptoms are similar to those of nitrogen deficiency – poor growth and yellow tinge to leaves. However, in sulphur deficiency, the youngest leaves are pale whereas in nitrogen deficiency the older leaves are most affected. There is an increasing need for the use of sulphur fertilisers in both arable and grassland systems (see page 43 for more details).

Sulphur recommendations are given as kg/ha of sulphur trioxide (SO_3) not as S. Conversion tables are given in Appendix 8.

Sodium

Sodium will not have any effect on grass growth but an adequate amount in the diet is essential for livestock health and can improve the palatability of grass. The sodium content of herbage is normally adequate for grazing livestock though it may be reduced if excess potash is applied. Herbage analysis is useful to assess the sodium status of grass. Where sodium levels are low, mineral supplements may be required for some classes of stock or a sodium containing fertiliser may be used.

Lime and micronutrients (trace elements)

Many grass species can tolerate more acid conditions than most arable crops, but grass/clover swards are less tolerant of soil acidity than all-grass swards. Clover is less likely to persist where the soil pH is below the optimum.

The optimum soil pH for continuous grassland is 6.0 (mineral soils) and 5.3 (peaty soils). In a mixed grass/arable rotation where an occasional cereal crop is grown in a predominantly grassland rotation, the soil pH of mineral soils should be maintained at 6.0 or 6.2 if barley is grown.

Over-liming should be avoided as it can induce deficiencies of trace elements such as copper, cobalt and selenium which can adversely affect livestock growth but will not affect grass growth. Where a deficiency does occur, treatment of the animal with the appropriate trace element is usually the most effective means of control, though application of cobalt and selenium to grazing pastures can be effective.

More information on the use of lime is given in Section 1.

Section 8: Grass

Protecting the Environment

Losses of nutrients from agricultural land can pollute both ground and surface waters and the atmosphere. It is important that farmers do everything possible to minimise the risk of agricultural practices causing environmental pollution.

Some areas of agricultural land are subject to various types of agreement for maintenance or improvement of the environment and farmers may have to comply with restrictions on the use of fertilisers and manures. Farmers in these areas may need to modify the recommendations in this Manual in order to comply with specific requirements.

There is guidance in *Protecting Our Water, Soil and Air: A Code of Good Agricultural Practice* and in the NVZ Guidance Leaflets (Section 9).

Environmental issues are described in Section 1. There are some issues particularly associated with grassland due to common conditions (sloping land, high rainfall, soils relatively high in organic matter) and to the presence of manures, either applied or deposited by grazing animals.

Nitrate

- Farmers of land within NVZs have to comply with mandatory rules which aim to reduce nitrate movement to water. These rules are set out in NVZ Guidance Leaflets (Section 9).
- Movement of nitrate to water can be reduced by careful matching of rate and timing of fertiliser nitrogen applications to the needs of the crop. Account must be taken of nitrogen supplied by the soil, previous crop residues and organic manures.
- Other measures which will reduce the risk of nitrate pollution include:
  - applying organic manures and fertilisers close to the time when the nitrogen is needed for crop growth.
  - avoiding autumn and early winter application of manures wherever practically possible.
  - reducing fertiliser nitrogen use where grass growth is restricted by summer drought.

Ammonia

- Ammonia deposition is damaging to the environment due to its contribution to acid rain and excessive enrichment of nutrient-poor habitats.
- Livestock manures are the largest source of ammonia with greatest losses from housing and spreading to land.
- Ammonia emissions can be reduced by incorporating manures after spreading (where possible) and by applying slurry with a band spreader or injector.
- Reducing ammonia loss from late winter and spring applications of manure will also conserve available nitrogen for crop uptake.

June 2010

Section 8: Grass

Phosphorus

- Phosphorus can be damaging to water quality in rivers and lakes where it may cause excessive algal or weed growth.
- Phosphorus is lost from grassland mainly in surface run-off from recently spread manures, soil erosion, or dissolution of readily-available soil P, and in drain outflows in soluble forms or attached to soil particles.
- The risk of phosphorus loss can be reduced by:
  - taking full account of the phosphate content of organic manures when deciding on fertiliser requirements.
  - following the recommendations in this Manual to avoid increasing soil P Indices beyond those necessary for crop production.
  - avoiding surface applications of manure when ground conditions are unsuitable or on steeply sloping land adjacent to water courses
  - reducing reliance on imported concentrates

Nitrous oxide

- Nitrous oxide, a powerful greenhouse gas, is formed during denitrification of nitrate-N and nitrification of ammonium-N in the soil.
- The amount emitted is related to concentrations of ammonium-N and of nitrate-N which can be restricted by matching nitrogen inputs to grass demand.
- Using recommendations in this Manual and, in particular, making effective use of nitrogen in organic manures will help minimise emission.

Section 8: Grass

Finding the Nitrogen Fertiliser Recommendation

Procedure

- Using the table on page 188, identify the Soil Nitrogen Supply status of the field using information for:
  - typical nitrogen use in the last 2-3 years
  - grass management last year
- Using the table on page 188, identify the Grass Growth Class of the field using information for:
  - the predominant soil type in the field (see page 17 and Appendix 1)
  - average summer rainfall
- Using the factors on page 189, calculate stocking rate in livestock units/ha (LU/ha).
- Decide on the intended grass management and select one of the recommendation tables on pages 191-213:
  - Grass Establishment
  - Dairy production
  - Beef production
  - Sheep production
  - Grass/clover swards
  - Hay
- During the season adjust nitrogen use according to the weather and actual grass growth performance (see page 189).

Assessing the Soil Nitrogen Supply status

The nitrogen recommendations are based on the requirement of the crop to be grown, making allowance for soil nitrogen residues. In grassland systems, these residues are assessed in a different way to that used for arable or vegetable cropping systems. Three levels of soil nitrogen status are recognised and used in the recommendation tables. Fields with a low soil nitrogen status need more nitrogen compared with fields with a moderate or high status.

Nitrogen fertiliser, organic manure use and the grass management history in the last 1-3 years is of most importance for determining the soil nitrogen status, but longer histories can be relevant. Applications of organic manures over the last few years must also be taken into account.

Section 8: Grass

Soil Nitrogen Supply (SNS) Status in Grassland Systems According to Previous Grass Management

	Previous grass management	Previous nitrogen use (kg/ha)[a]
High	Long term grass, high input. Includes: • grass reseeded after grass or after 1 year arable • grass ley in second or later year	over 250
Moderate[b]	First year ley after 2 or more years arable (last crop potatoes, oilseed rape, peas or beans, NOT on light sand soil) Long term grass, moderate input. Includes: • grass reseeded after grass or after 1 year arable • grass ley in second or later year	All 100 – 250 or Substantial clover content
Low	First year ley after 2 or more years arable (last crop cereal, sugar beet, linseed or any crop on a light sand soil) Long term grass, low input. Includes: • grass reseeded after grass or after 1 year arable • grass ley in second or later year	All Up to 100

a. Refers to typical fertiliser and available manure nitrogen used per year in the last 2-3 years.

b. The nitrogen values in the recommendation tables assume a moderate Soil Nitrogen Supply status and so adjustments need to be made only for high or low Soil Nitrogen Supply: increase total fertiliser nitrogen input by 30 kg/ha in a low SNS situation; decrease total fertiliser nitrogen input by 30 kg/ha in a high SNS situation.

- Increase the soil nitrogen status by one class if more than 150 kg/ha of total nitrogen has been regularly applied as organic manure for several years. Reduce the soil nitrogen status by one class if grass was cut for silage and less than 150 kg/ha of total nitrogen as organic manure has been applied on average in previous years.

Assessing Grass Growth Class

Grass growth will be restricted where summer rainfall plus the moisture stored in the soil (the soil available water) are inadequate to meet the grass demand for water. Although there can be wide variations in summer rainfall between years, the table below gives an indication of the grass growth potential **in an average season**, based on the risk of summer drought. For simplicity within the recommendation tables, we combine the 'very poor' and 'poor' grass growth classes. We also combine the 'very good' and 'good' grass growth classes.

Grass Growth Classes

Soil Available Water	Soil types[a]	Rainfall[b] (April to September inclusive)		
		up to 300 mm	300 – 400 mm	over 400 mm
Low	Light sand soils and shallow soils (not over chalk)	Very poor	Poor	Average
Medium	Medium soils, deep clay soils, and shallow soils over chalk	Poor	Average	Good
High	Deep silty soils, peaty soils and soils with groundwater (e.g. river meadows)	Average	Good	Very good

a. See Appendix 1 for soil descriptions.

b. Mean summer rainfall (April to September) is usually about half of annual rainfall.

Section 8: Grass

For sites above 300 m altitude, reduce the growth class by one. This is because lower temperatures will restrict growth.

The recommendation tables indicate how it is possible to decrease total nitrogen requirements the better the Grass Growth Class in order to produce the target amount of home grown forage, as nitrogen is used more efficiently. In other words, the better the Grass Growth Class, the greater the efficiency of nitrogen use and a reduced risk of nitrogen losses to the environment. Conversely, the poorer the Grass Growth Class the higher the nitrogen requirement (assuming no other limiting factors, such as shortage of phosphate, potash and sulphur supply). This reflects the need to support a target economic level of animal production rather than maximum yield of forage.

Good nutrient management is important to all farming systems and at all levels of nutrient application. In some situations where higher nitrogen requirements are advised, carefully planned use of fertiliser will maximise the nutrients taken up by the crop rather than being lost to the environment. Where good growth is expected lower nitrogen requirements are possible and may be advised, and in this situation good nutrient management is important to ensure nutrients are used efficiently in order to maintain yield.

Calculating stocking rate

Stocking rate is one important factor in determining requirement for herbage. It is usually expressed in livestock units/ha (LU/ha). In estimating the LU for N recommendations: 1 (650 kg) cow = 1LU (other stock can be expressed as LU by pro rata on liveweight, e.g. calves), 1 average beef animal = 0.6 LU and 1 ewe with lambs = 0.17 LU. Calculate the average number of livestock as LU and divide by the total area of grassland in the enterprise to give stocking rate.

Adjusting Nitrogen Use for Weather during the Growing Season

If grass yields are restricted due to drought, reduce the use of nitrogen once growth restarts following rain. As a guide, if grass does not grow for about 2 weeks in June or July, the yield will be reduced by about 1 t/ha of dry matter and there will be about 40 kg/ha of unused nitrogen remaining in the soil. This residual nitrogen must be allowed for when deciding on nitrogen use once grass growth starts again following rain.

Section 8: Grass

Finding the Phosphate, Potash and Magnesium Recommendations

For phosphate, potash and magnesium, the results of a recent soil analysis will be needed showing the soil Index. The use of soil analysis as a basis for making fertiliser decisions is described in Section 1, and the procedure for taking soil samples in Appendix 3. Soil samples should be taken from every field every 3-5 years.

Recommendations are given in the tables as phosphate (P_2O_5), potash (K_2O) and magnesium oxide (MgO). Conversion tables are given in Appendix 8.

- Recommendations at target Indices (2 for P and 2- for K) are maintenance dressings intended to maintain soil reserves and prevent depletion of soil fertility rather than give a yield response. Recommendations at Indices lower than target include an allowance for building-up soil reserves over several years as well as meeting immediate crop requirement.
- All recommendations are given for the mid-point of each Index (mid-point of 2- for potash).
- Phosphate and potash recommendations are given for crops producing average yields. The recommendation may be increased or decreased where yields are substantially more or less than this yield level. These adjustments can be calculated using the values for phosphate and potash in grass (see Appendix 5).

> ### Example 1
> *Soil analysis shows P Index 3 and K Index 2-. A pure grass sward is to be closed up for 2 cuts of silage. No organic manures will be applied.*
>
> The table on page 211 recommends 20 kg P_2O_5/ha and 80 kg K_2O/ha for the first cut, and nil phosphate and 90 kg K_2O/ha immediately after the first cut for second cut growth. A further 80 kg K_2O/ha should be applied after the second cut – see notes beneath the table.

Section 8: Grass

Grass Establishment – Nitrogen

	Soil Nitrogen Supply		
	Low	Moderate	High
	kg/ha		
Autumn sown	0	0	0
Spring sown	60	60	60
Grass/clover swards	0	0	0

Grass Establishment – Phosphate, Potash And Magnesium

	P or K Index				
	0	1	2	3	Over 3
	kg/ha				
Phosphate (P_2O_5)	120	80	50	30	0
Potash (K_2O)	120	80	60 (2-) 40 (2+)	0	0

The amount of phosphate and potash applied for establishment may be deducted from the first season's grazing or silage/hay requirement.

For magnesium recommendations, see page 183.

Don't forget to deduct nutrients applied as organic manures – see Section 2

Section 8: Grass

Dairy Production – Nitrogen Requirements

The recommendations for promoting the growth of grass for animal production are based on knowledge of grass response to fertiliser nitrogen, under conditions where growth is not limited by supplies of other nutrients. The recommendations are based on the need to produce the amount of home grown forage necessary to maintain a target intensity of production, rather than the optimal amount relative to the cost of fertiliser. This enables farmers who may be operating at widely different stocking rates and feeding different levels of concentrates to obtain relevant recommendations for whole season nitrogen requirements. Nitrogen requirements have been calculated using commonly accepted values for feed energy conversion to milk, proportions of available land devoted to cutting and grazing and efficiencies with which home-grown forage can be utilised under the two harvesting regimes (cut and grazed grass).

The N recommendations for dairy production are based on feeding a herd of Holstein-Friesian animals (where 1 adult = 650 kg liveweight = 1 LU) producing standard milk composition (**4.0% fat and 3.3% protein**). Concentrate use ranges from 0.5t to 4.4t/cow/yr. It is assumed that 40% of the forage DM comes from cut grass, the remainder from grazed grass, and that silage production is optimised, whatever the overall level of production, resulting in (in general) higher levels of N requirement under cutting than grazing. A 3-cut system is assumed – but advice is provided about how to operate with fewer cuts. Note, the whole season total N requirement can be provided through contributions from SNS, clover and applied organic manures, as well as fertiliser. If you operate a markedly different system to this, you may need to seek FACTS qualified advice about interpretation of the N recommendations.

Three Grass Growth Classes have been defined earlier (see page 188): *Very good/Good*, *Average* and *Poor/Very poor*. Their definition depends on the water holding capacity of the soil and amount of summer rainfall.

There are two tables for dairy production; one for nitrogen requirements for cutting and grazing on Grass Growth Class *Very Good/Good* (Table 8.1) and a second table for nitrogen requirements for cutting and grazing on *Average* Grass Growth Class (Table 8.2). Therefore, the first step is to identify the Grass Growth Class of your land to direct you to the correct table. It is assumed that dairy farming requires a Grass Growth Class of at least *Average*. Within each table there are total whole season nitrogen requirement values given, according to level of production, concentrate use and stocking rate to enable most of the systems operating in UK grassland farming to be matched as closely as possible.

Adjustments can be made (indicated above the tables) for high or low soil nitrogen supply (SNS – see page 188), and for nitrogen from applied manures. Suggestions for the percentage splits and their timing for each total fertiliser amount are also given within the tables and accompanying notes.

To use the recommendations, first identify the Grass Growth Class of your land to direct you to the correct table. Within each of the dairy tables the system is first fixed by the average annual milk yield per cow. Each of these milk yields is then subdivided according to concentrate

Section 8: Grass

use and each of these further subdivided according to stocking rate. Thus, systems are grouped in sets of three according to milk yield and concentrate use with a range of stocking rates. The maximum stocking rate which can be supported for each milk yield/concentrate combination is given in the top line of each set of three.

First Actions:

- Identify whether your land is Very *good/Good* or *Average* Grass Growth Class (page 188)
- Choose Table 8.1 (*Good/Very good*) or Table 8.2 (*Average*)
- Identify the SNS of the land (Low, Moderate, High) (page 188)
- Find the milk yield/concentrate/stocking rate division most appropriate to your management.

Table 8.1 provides the whole season total nitrogen requirements for cutting and grazing in the *Good/Very good* Grass Growth Class (most dairy enterprises). Values in the table are for a moderate SNS. **Reduce total fertiliser nitrogen input by 30 kg N/ha for high SNS. Increase total fertiliser nitrogen input by 30 kg N/ha for low SNS (uncommon for dairy production).**

Section 8: Grass

Table 8.1 Dairy: Grass Growth Class *Very Good / Good*. Whole-season total nitrogen requirement for cut and grazed grass (kg N/ha).

	DAIRY Grass Growth Class *Very Good / Good*				
			Total N Requirement		
Milk yield	**Concentrate use**	**Stocking Rate**	**Cut**		**Grazed**
l/cow/yr	t/cow/y	LU/ha	kg/ha	Indicative yield* (t DM/ha)	kg/ha
8,000 to 10,000	4.4	4.0	360	11.2	280
		3.5	310	10.6	210
		3.0	260	9.7	150
8,000 to 10,000	3.7	3.0	370	11.3	300
		2.6	310	10.5	210
		2.2	240	9.4	150
6,000 to 8,000	2.2	2.6	360	11.2	320
		2.2	280	10.1	210
		1.8	200	8.5	140
6,000 to 8,000	1.5	2.2	360	11.2	340
		1.8	260	9.8	210
		1.6	210	8.8	170
6,000 to 8,000	0.9	1.9	330	10.9	320
		1.7	280	10.1	240
		1.5	230	9.1	190
4,000 to 6,000	0.9	2.4	350	11.0	330
		2.1	280	10.1	240
		1.8	210	8.8	180
4,000 to 6,000	0.5	2.1	320	10.7	290
		1.9	270	9.9	230
		1.7	220	8.9	190
< 5,000 extended grazing	0.5	2.5	350	11.1	340
		2.2	310	10.6	240
		2.0	270	9.9	200

Values in the above table are total annual nitrogen requirement. To obtain your fertiliser requirement, DON'T FORGET to consider nitrogen supply from SNS (page 188), clover (page 208) and applied organic manures (Section 2).

*Average dry matter yields for cut grass should be achievable in most seasons.

- concentrate use is based on that used per lactating cow over a calendar year
- stocking rate is based on all dairy animals (lactating cows plus followers), 1 cow = 1 LU
- To interpolate between **stocking rates** at a given milk yield/concentrate use combination, assume a proportional difference in nitrogen requirement between two values.

Section 8: Grass

- To interpolate between **concentrate use** within a given milk yield/stocking rate combination:
  - **add** 10 kg N/ha for every reduction of 0.1 tonnes concentrate/year/animal
  - **subtract** 10 kg N/ha for every increase of 0.1 tonnes concentrate/year/animal.
- Seek additional help from a FACTS qualified advisor if you have difficulty in matching your system to those in the table.

The total nitrogen requirement should be split into 3-6 applications during the growing season if > 150 kg N/ha. Three applications would be appropriate only where the total requirement is less than/equal to150 kg N/ha. Applications need not be of equal size and advantage can be taken of the relatively high grass growth rate in late spring. For example, the recommended split of N application could be; for **first cut**; 40% (could be split further, Feb-March 15%, April 25%), for **second cut**; 35% (could be split further, May 20%, June 15%); for **subsequent cuts**; 25% (could be split further, July 15%, August 10%).

In mild areas where earlier grazing is possible, nitrogen may be applied from early-mid February. In upland areas, apply nitrogen from mid-late March. Typically, nitrogen would be applied around one month before normal turn-out date.

Cutting after early spring grazing

Following early spring grazing, reduce the 1st cut recommendation by 25 kg/ha. For grazing after cutting, see recommendations in the *Grazing and cutting* section later in the dairy section.

Table 8.2 provides the whole season total nitrogen requirements for cutting and grazing in the *Average* Grass Growth Class. Values in the table are for a moderate SNS situation. **Reduce total fertiliser nitrogen input by 30 kg N/ha in a high SNS situation. Increase total fertiliser nitrogen input by 30 kg N/ha in a low SNS situation (uncommon for dairy production).**

Section 8: Grass

Table 8.2 Dairy: Grass Growth Class *Average*. Whole-season total nitrogen requirement for cut and grazed grass (kg N/ha).

			DAIRY Grass Growth Class *Average*		
			Total N Requirement		
Milk yield	**Concentrate use**	**Stocking Rate**	**Cut**		**Grazed**
l/cow/yr	t/cow/y	LU/ha	kg/ha	Indicative yield* (t DM/ha)	kg/ha
8,000 to 10,000	4.4	3.5	340	9.7	290
		3.1	300	9.3	210
		2.8	260	8.8	170
8,000 to 10,000	3.7	2.6	330	9.7	290
		2.4	300	9.3	240
		2.2	270	8.8	200
6,000 to 8,000	2.2	2.3	320	9.6	340
		2.1	290	9.1	260
		1.8	230	8.2	180
6,000 to 8,000	1.5	1.9	310	9.4	340
		1.7	260	8.8	240
		1.5	220	8.0	190
6,000 to 8,000	0.9	1.7	300	9.4	340
		1.5	250	8.6	240
		1.3	200	7.6	180
4,000 to 6,000	0.9	2.1	300	9.4	330
		1.9	260	8.8	250
		1.7	220	8.0	200
4,000 to 6,000	0.5	1.9	290	9.2	330
		1.7	250	8.5	240
		1.5	200	7.7	190
< 5,000 extended grazing	0.5	2.2	340	9.7	340
		2	300	9.3	250
		1.8	270	8.9	200

- Values in the above table are total annual nitrogen requirement. To obtain your fertiliser requirement, DON'T FORGET to consider nitrogen supply from SNS (page 188), clover (page 208) and applied organic manures (Section 2)
- *Average dry matter yields for cut grass should be achievable in most seasons.
- concentrate use is based on that used per lactating cow over a calendar year
- stocking rate is based on all dairy animals (lactating cows plus followers), 1 cow = 1 LU
- To interpolate between **stocking rates** at a given milk yield/concentrate use combination, assume a proportional difference in nitrogen requirement between two values.

Section 8: Grass

- To interpolate between **concentrate use** within a given milk yield/stocking rate combination:
  - **add** 10 kg N/ha for every reduction of 0.1 tonnes concentrate/year/animal
  - **subtract** 10 kg N/ha for every increase of 0.1 tonnes concentrate/year/animal.
- Seek additional help from a FACTS qualified advisor if you have difficulty in matching your system to those in the table.

The total nitrogen requirement should be split into 3-6 applications during the growing season if > 150 kg N/ha. Three applications would be appropriate only where the total requirement is less than/equal to150 kg N/ha. Applications need not be of equal size and advantage can be taken of the relatively high grass growth rate in late spring. For example, the recommended split of N application could be; for **first cut**; 40% (could be split further, Feb-March 15%, April 25%), for **second cut**; 35% (could be split further, May 20%, June 15%); for **subsequent cuts**; 25% (could be split further, July 15%, August 10%).

In mild areas where earlier grazing is possible, nitrogen may be applied from early-mid February. In upland areas, apply nitrogen from mid-late March. Typically, nitrogen would be applied around one month before normal turn-out date.

Cutting after early spring grazing

Following early spring grazing, reduce the 1st cut recommendation by 25 kg/ha. For grazing after cutting, see recommendations in the *Grazing and cutting* see below.

Grazing and cutting systems

Where cut grass is followed by grazing, you should obtain the two total N requirement values (cut and grazed) for your chosen system. For cut grass, the recommended split of N application could be; for **first cut**; 40% (could be split further, Feb-March 15%, April 25%), for **second cut**; 35% (could be further split May 20%, June 15%); for **subsequent cuts** 25% (could be further split July 15%, August 10%). For grazed grass the recommended split could be: Feb-March 15%, April 25%, May 20%, June 15%; July 15%, August 10%.

Then, for example in the case of a 1 cut system followed by grazing; follow the guidance on proportional total N split for 1st cut, i.e. 40% of the total value. Then revert to 20% proportional split of grazed grass total value (starting May) and follow grazing split accordingly.

Similarly, for a 2 cut system followed by grazing; follow the guidance on the proportional total N split for 1st and 2nd cuts, i.e. 40% for 1st cut and 35% for 2nd cut. Then revert to the 15% proportion split of the grazed grass total value (starting July) and follow grazing split accordingly.

Section 8: Grass

Beef Production – Nitrogen Requirements

The recommendations for promoting the growth of grass for animal production are based on knowledge of grass response to fertiliser nitrogen, under conditions where growth is not limited by supplies of other nutrients. The recommendations are based on the need to produce the amount of home grown forage necessary to maintain a target intensity of production, rather than the optimal amount relative to the cost of fertiliser. This enables farmers who may be operating at widely different stocking rates and feeding different levels of concentrates to obtain relevant recommendations for whole season nitrogen requirements. Nitrogen requirements have been calculated using commonly accepted values for feed conversion to liveweight gain, proportions of available land devoted to cutting and grazing and efficiencies with which home-grown forage can be utilised under the two harvesting regimes (cutting and grazing).

The N recommendations for beef production are based on targets of achieving set liveweight gains using standard feed conversion factors for home grown forage and concentrates. The intensively and moderately grazed beef systems are based on a typical housing period of 170d, with concentrate use of 0.2 to 0.4t/animal/yr, fed through the **winter** housing season. The extensively grazed system assumes 110d housing period and no concentrate use. Grazed and cut grass provide the remaining energy to maintain target livestock growth rates of 0.95 kg/head/d (intensively grazed), 0.85 kg/head/d (moderately grazed) and 0.6 kg/head/d (extensive grazing). Silage production is optimised, whatever the overall level of production, resulting in (in general) higher levels of N requirement under cutting than grazing. A 3-cut system is assumed – but advice is provided about how to operate with fewer cuts. The systems approach assumes an operational livestock unit conversion factor for beef production of 0.6 (rather than 0.75 in other literature) as this better represents the average animal size across a typical beef farm. Note, the whole season total N requirement can be provided through contributions from SNS, clover and applied organic manures, as well as fertiliser. If your system is markedly different from this, then you may need to seek FACTS qualified advice about interpretation of N recommendations.

Three Grass Growth Classes have been defined earlier (see page 188): *Very good/Good*, Average and *Poor/Very poor*. Their definition depends on the water holding capacity of the soil and amount of summer rainfall.

There are three tables for beef production; one for nitrogen requirements for cutting and grazing for Grass Growth Class *Very good/Good* (Table 8.3), one for the *Average* Grass Growth Class (Table 8.4), and one for the *Poor/Very poor* Grass Growth Class (Table 8.5). Therefore the first step is to identify the Grass Growth Class of your land to direct you to the correct table. Within each table there are total nitrogen requirement values given for three different levels of production intensity according to concentrate use and stocking rate to enable most of the systems operating in UK grassland farming to be matched as closely as possible.

Adjustments can be made (indicated above the tables) for high or low soil nitrogen supply (SNS – see page 188), and for nitrogen from applied manures. Suggestions for the percentage splits and their timing for each total fertiliser amount are also given within the tables and accompanying notes.

To use the recommendations, first identify the Grass Growth Class of your land to direct you to the correct table. Within the beef tables some intensive, moderate and extensive systems are identified, which are then each subdivided according to concentrate use, followed by stocking rate.

Section 8: Grass

First Actions:

- Identify whether your land is *Very good/Good, Average* or *Poor/Very poor* Grass Growth Class (page 188).
- Choose Table 8.3 (*Good/Very good*), Table 8.4 (*Average*) or Table 8.5 (*Poor/Very poor*).
- Identify the SNS of the land (Low Moderate, High) (page 188).
- Find the concentrate/stocking rate division most appropriate to your management.

Table 8.3 provides the whole season total nitrogen requirements for cutting and grazing in the *Very good/Good* Grass Growth Class. Values in the table are for a moderate SNS situation. **Reduce total fertiliser nitrogen input by 30 kg N/ha for high SNS. Increase total fertiliser nitrogen input by 30 kg N/ha for low SNS.**

Table 8.3 Beef: Grass Growth Class *Very good/Good*. Whole-season total nitrogen requirement for cut and grazed grass (kg N/ha)

BEEF Grass Growth Class *Very good/Good*			Total N requirement		
	Concentrate use	Stocking Rate	Cut		Grazed
	(t/animal/year)	LU/ha	kg/ha	Indicative yield*(t DM/ha)	kg/ha
Intensively grazed (lowland dairy steers and heifers; some suckler herds)	0.4	2.2	370	11.3	330
		1.9	350	11.0	220
		1.6	320	10.7	150
Moderately grazed (upland and lowland suckler herds; lowland dairy steers and heifers)	0.2	1.4	330	10.8	110
		1.2	300	10.5	70
		1.0	280	10.2	30
Extensively grazed (moorland/hill beef; grazing for biodiversity)	0.0	0.9	240	9.4	0
		0.5	220	8.8	0
		0.3	200	8.5	0

- Values in the above table are total annual nitrogen requirement. To obtain your fertiliser requirement, DON'T FORGET to consider nitrogen supply from SNS (page 188), clover (page 208) and applied organic manures (Section 2)
- *Average dry matter yields for cut grass should be achievable in most seasons.
- For estimation of LU for beef systems, use an average of 0.6 LU (= 360kg) per animal
- For stocking rates that are intermediate between values in the table, assume a proportional difference in nitrogen requirement between two values.

Section 8: Grass

The total nitrogen requirement should be split into 3-6 applications during the growing season if > 150 kg N/ha. Three applications would be appropriate only where the total requirement is less than or equal to 150 kg N/ha. Applications need not be of equal size and advantage can be taken of the relatively high grass growth rate in late spring. For example, the recommended split of N application could be; for **first cut**; 40% (could be split further, Feb-March 15%, April 25%), for **second cut**; 35% (could be split further, May 20%, June 15%); for **subsequent cuts**; 25% (could be split further, July 15%, August 10%).

In mild areas where earlier grazing is possible, nitrogen may be applied from early-mid February. In upland areas, apply nitrogen from mid-late March. Typically, nitrogen would be applied around one month before normal turn-out date.

Cutting after early spring grazing

Following early spring grazing, reduce the 1st cut recommendation by 25 kg/ha. For grazing after cutting, see recommendations in the *Grazing and cutting* section later in the beef section.

Table 8.4 provides the whole season total nitrogen requirements for cutting and grazing in the *Average* Grass Growth Class. Values in the table are for a moderate SNS situation. **Reduce total fertiliser nitrogen input by 30 kg N/ha for high SNS. Increase total fertiliser nitrogen input by 30 kg N/ha for low SNS.**

Table 8.4 Beef: Grass Growth Class *Average*. Whole-season total nitrogen requirement for cut and grazed grass (kg N/ha)

BEEF Grass Growth Class *Average*				Total N requirement		
	Concentrate use	Stocking Rate		Cut		Grazed
	(t/animal/year)	LU/ha	kg/ha	Indicative yield* (t DM/ha)		kg/ha
Intensively grazed (lowland dairy steers and heifers; some suckler herds)	0.4	2.0	340	9.8		330
		1.8	320	9.6		240
		1.6	300	9.4		180
Moderately grazed (upland and lowland suckler herds; lowland dairy steers and heifers)	0.2	1.4	310	9.5		140
		1.2	290	9.2		90
		1.0	280	9.0		40
Extensively grazed (moorland/hill beef; grazing for biodiversity)	0.0	0.9	240	8.3		0
		0.5	210	7.9		0
		0.3	200	7.6		0

- Values in the above table are total annual nitrogen requirement. To obtain your fertiliser requirement, DON'T FORGET to consider nitrogen supply from SNS (page 188), clover (page 208) and applied organic manures (Section 2)
- *Average dry matter yields for cut grass should be achievable in most seasons.
- For estimation of LU for beef systems, use an average of 0.6 LU (= 360kg) per animal.
- For stocking rates that are intermediate between values in the table, assume a proportional difference in nitrogen requirement between two values.

Section 8: Grass

The total nitrogen requirement should be split into 3-6 applications during the growing season if > 150 kg N/ha. Three applications would be appropriate only where the total requirement is less than or equal to 150 kg N/ha. Applications need not be of equal size and advantage can be taken of the relatively high grass growth rate in late spring. For example, the recommended split of N application could be; for **first cut**; 40% (could be split further, Feb-March 15%, April 25%), for **second cut**; 35% (could be split further, May 20%, June 15%); for **subsequent cuts**; 25% (could be split further, July 15%, August 10%).

In mild areas where earlier grazing is possible, nitrogen may be applied from early-mid February. In upland areas, apply nitrogen from mid-late March. Typically, nitrogen would be applied around one month before normal turn-out date.

Cutting after early spring grazing

Following early spring grazing, reduce the 1st cut recommendation by 25 kg/ha. For grazing after cutting, see recommendations in the *Grazing and cutting* section later in the beef section.

Table 8.5 provides the whole season total nitrogen requirements for cutting and grazing in the *Very poor/Poor* Grass Growth Class. Values in the table are for a moderate SNS situation. **Reduce total fertiliser nitrogen input by 30 kg N/ha for high SNS. Increase total fertiliser nitrogen input by 30 kg N/ha for low SNS.**

Table 8.5 Beef: Grass Growth Class *Very poor/Poor*. Whole-season total nitrogen requirement for cut and grazed grass (kg N/ha)

BEEF Growth Class *Very poor/Poor*				Total N requirement	
	Concentrate use	Stocking rate		Cut	Grazed
	(t/animal/year)	LU/ha	kg/ha	Indicative yield* (t DM/ha)	kg/ha
Intensively grazed (lowland dairy steers and heifers; some suckler herds)	0.4	1.6	280	7.0	250
		1.4	270	6.9	170
		1.2	260	6.7	110
Moderately grazed (upland and lowland suckler herds; lowland dairy steers and heifers)	0.2	1.4	290	7.0	200
		1.2	270	6.9	120
		1.0	260	6.7	60
Extensively grazed (moorland/hill beef; grazing for biodiversity)	0.0	0.9	230	6.3	30
		0.5	210	5.9	0
		0.3	200	5.8	0

- Values in the above table are total annual nitrogen requirement. To obtain your fertiliser requirement, DON'T FORGET to consider nitrogen supply from SNS (page 188), clover (page 208) and applied organic manures (Section 2)
- *Average dry matter yields for cut grass should be achievable in most seasons.
- For estimation of LU for beef systems, use an average of 0.6 LU (= 360kg) per animal.
- For stocking rates that are intermediate between values in the table, assume a proportional difference in nitrogen requirement between two values.

Section 8: Grass

The total nitrogen requirement should be split into 3-6 applications during the growing season if > 150 kg N/ha. Three applications would be appropriate only where the total requirement is less than or equal to 150 kg N/ha. Applications need not be of equal size and advantage can be taken of the relatively high grass growth rate in late spring. For example, the recommended split of N application could be; for **first cut**; 40% (could be split further, Feb-March 15%, April 25%), for **second cut**; 35% (could be split further, May 20%, June 15%); for **subsequent cuts**; 25% (could be split further, July 15%, August 10%).

In mild areas where earlier grazing is possible, nitrogen may be applied from early-mid February. In upland areas, apply nitrogen from mid-late March. Typically, nitrogen would be applied around one month before normal turn-out date.

Cutting after early spring grazing

Following early spring grazing, reduce the 1st cut recommendation by 25 kg/ha. For grazing after cutting, see recommendations in the *Grazing and cutting* section below.

In paddock systems, nitrogen should be applied within 48 hours after grazing as further delay will reduce grass yield.

In mild areas where earlier grazing is possible, nitrogen may be applied from early-mid February. In upland areas, apply nitrogen from mid-late March.

Application of nitrogen usually is not justified after mid-August in intensive grazing systems as grass response declines and there is increased risk of nitrate leaching.

Grazing and cutting systems

Where cut grass is followed by grazing, farmers should obtain the two total N requirement values (cut and grazed) for their chosen system. For cut grass, the recommended split of N application could be; For first cut; 40% (could be split further, Feb-March 15%, April 25%), For second cut; 35% (could be further split May 20%, June 15%); subsequent cuts 25% (could be further split July 15%, August 10%). For grazed grass the recommended split could be: Feb-March 15%, April 25%, May 20%, June 15%; July 15%, August 10%.

Then, for example in the case of a 1 cut system followed by grazing; follow the guidance on proportional total N split for 1st cut, i.e. 40% of the total value. Then revert to 20% proportional split of grazed grass total value (starting May) and follow grazing split accordingly.

Similarly, for a 2 cut system followed by grazing; follow the guidance on the proportional total N split for 1st and 2nd cuts, i.e. 40% for 1st cut and 35% for 2nd cut. Then revert to the 15% proportion split of the grazed grass total value (starting July) and follow grazing split accordingly.

Section 8: Grass

Sheep Production – Nitrogen Requirements

The recommendations for promoting the growth of grass for animal production are based on knowledge of grass response to fertiliser nitrogen, under conditions where growth is not limited by supplies of other nutrients. The recommendations are based on the need to produce the amount of home grown forage necessary to maintain a target intensity of production, rather than the optimal amount relative to the cost of fertiliser. This enables farmers who may be operating at widely different stocking rates and feeding different levels of concentrates to obtain relevant recommendations for whole season nitrogen requirements. Nitrogen requirements have been calculated using commonly accepted values for feed intake, liveweight gain, proportions of available land devoted to cutting and grazing and efficiencies with which home-grown forage can be utilised under the two harvesting regimes (cutting and grazing). If hay is used to supplement feeding in the winter months, refer to the Table on page 213 for recommendations on nitrogen requirement.

The N recommendations for sheep are based on supplying sufficient home grown forage and concentrate to maintain target annual intake at 3 levels of production. Ewes are fed concentrates during lambing in the intensively and moderately grazed systems (1.0 and 0.5 kg/ewe/d, respectively). They also receive conserved forage at this time. Lambs may also need to be fed concentrates to maintain the higher stocking rates. Assumptions regarding LU are as follows: 1 ewe + lambs = 0.17 LU in the intensively and moderately grazed systems, and 0.14 LU in the extensively grazed system. As with dairy and beef, silage production is optimised at relatively high levels of N requirement, and hence in the moderately grazed system there is a significant N requirement for cut grass, and little or none for grazed grass. Note, the whole season total N requirement can be provided through contributions from SNS, clover and applied organic manures, as well as fertiliser. If your system is markedly different from this, then you may need to seek FACTS qualified advice about interpretation of N recommendations.

Three Grass Growth Classes have been defined earlier (see page 188): *Very good/Good*, *Average* and *Poor/Very poor*. Their definition depends on the water holding capacity of the soil and amount of summer rainfall.

There are three tables for sheep production; one for nitrogen requirements for cutting and grazing for Grass Growth Class *Very good/Good* (Table 8.6), one for the *Average* Grass Growth Class (Table 8.7), and one for the *Poor/Very poor* Grass Growth Class (Table 8.8). Therefore the first step is to identify the Grass Growth Class of your land to direct you to the correct table. Within each table there are total nitrogen requirement values given for three different levels of production intensity according to stocking rate to enable most of the systems operating in UK grassland farming to be matched as closely as possible.

Adjustments can be made (indicated above the tables) for high or low soil nitrogen supply (SNS – see page 188), and for nitrogen from applied manures. Suggestions for the percentage splits and their timing for each total fertiliser amount are also given within the tables and accompanying notes.

To use the recommendations, first identify the Grass Growth Class of your land to direct you to the correct table. Within the sheep tables some intensive, moderate and extensive systems are identified, which are then each subdivided according to stocking rate.

Section 8: Grass

First Actions:

- Identify whether your land is *Very good/Good, Average* or *Poor/Very poor* Grass Growth Class (page 188)
- Choose Table 8.6 (*Good/Very good*), Table 8.7 (*Average*) or Table 8.8 (*Poor/Very poor*)
- Identify the SNS of the land (Low Moderate, High) (page 188)
- Find the stocking rate division most appropriate to your management

Table 8.6 provides the whole season total nitrogen requirements for cutting and grazing in the Very good/Good Grass Growth Class. Values in the table are for a moderate SNS situation. **Reduce total fertiliser nitrogen input by 30 kg N/ha for high SNS. Increase total fertiliser nitrogen input by 30 kg N/ha for low SNS (uncommon for dairy production).**

Table 8.6 Sheep: Grass Growth Class *Very good/Good*. Whole-season total nitrogen requirement for cut and grazed grass (kg N/ha).

SHEEP Grass Growth Class *Very good/Good*	Stocking Rate	Total N requirement		Grazed
		Cut		
	LU/ha	kg/ha	Indicative yield* (t DM/ha)	kg/ha
Moderately to Intensively grazed	2.1	370	11.3	280
	1.7	340	10.9	170
	1.2	300	10.3	70
Extensively to Moderately grazed	1.0	280	10.1	40
	0.7	260	9.6	0
	0.5	240	9.3	0
Extensively grazed	0.2	210	8.9	0
	0.1	200	8.8	0
	0.05	200	8.5	0

- Values in the above table are total annual nitrogen requirement. To obtain your fertiliser requirement, DON'T FORGET to consider nitrogen supply from SNS (page 188), clover (page 208) and applied organic manures (Section 2)
- *Average dry matter yields for cut grass should be achievable in most seasons.
- For estimation of LU for sheep systems, use an average of 0.17 LU per ewe plus lambs. Stocking rates >2.0 increase the risk of high worm burden with consequences for animal productivity.
- For stocking rates that are intermediate between values in the table, assume a proportional difference in nitrogen requirement between two values.

Section 8: Grass

The total nitrogen requirement should be split into 3-6 applications during the growing season if > 150 kg N/ha. Three applications would be appropriate only where the total requirement is less than or equal to 150 kg N/ha. Applications need not be of equal size and advantage can be taken of the relatively high grass growth rate in late spring. For example, the recommended split of N application could be; for **first cut**; 40% (could be split further, Feb-March 15%, April 25%), for **second cut**; 35% (could be split further, May 20%, June 15%); for **subsequent cuts**; 25% (could be split further, July 15%, August 10%).

In mild areas where earlier grazing is possible, nitrogen may be applied from early-mid February. In upland areas, apply nitrogen from mid-late March. Typically, nitrogen would be applied around one month before normal turn-out date.

Cutting after early spring grazing

Following early spring grazing, reduce the 1st cut recommendation by 25 kg/ha. For grazing after cutting, see recommendations in the *Grazing and cutting* section later in the sheep section.

Table 8.7 provides the whole season total nitrogen requirements for cutting and grazing in the *Average* Grass Growth Class. Values in the table are for a moderate SNS situation. **Reduce total fertiliser nitrogen input by 30 kg N/ha for high SNS. Increase total fertiliser nitrogen input by 30 kg N/ha for low SNS (uncommon for dairy production).**

Table 8.7 Sheep: Grass Growth Class *Average*. Whole-season total nitrogen requirement for cut and grazed grass (kg N/ha).

SHEEP Grass Growth Class *Average*		Total N requirement		
	Stocking Rate	Cut		Grazed
	LU/ha	kg/ha	Indicative yield* (t DM/ha)	kg/ha
Moderately to Intensively grazed	2.0	340	9.7	350
	1.5	300	9.3	170
	1.0	270	8.8	50
Extensively to Moderately grazed	0.9	260	8.8	30
	0.7	250	8.6	0
	0.5	230	8.3	0
Extensively grazed	0.2	210	7.9	0
	0.1	200	7.7	0
	0.05	200	7.6	0

- Values in the above table are total annual nitrogen requirement. To obtain your fertiliser requirement, DON'T FORGET to consider nitrogen supply from SNS (page 188), clover (page 208) and applied organic manures (Section 2)
- *Average dry matter yields for cut grass should be achievable in most seasons.
- For estimation of LU for sheep systems, use an average of 0.17 LU per ewe plus lambs. Stocking rates >2.0 increase the risk of high worm burden with consequences for animal productivity.
- For stocking rates that are intermediate between values in the table, assume a proportional difference in nitrogen requirement between two values.

Section 8: Grass

The total nitrogen requirement should be split into 3-6 applications during the growing season if > 150 kg N/ha. Three applications would be appropriate only where the total requirement is less than or equal to 150 kg N/ha. Applications need not be of equal size and advantage can be taken of the relatively high grass growth rate in late spring. For example, the recommended split of N application could be; for **first cut**; 40% (could be split further, Feb-March 15%, April 25%), for **second cut**; 35% (could be split further, May 20%, June 15%); for **subsequent cuts**; 25% (could be split further, July 15%, August 10%).

In mild areas where earlier grazing is possible, nitrogen may be applied from early-mid February. In upland areas, apply nitrogen from mid-late March. Typically, nitrogen would be applied around one month before normal turn-out date.

Cutting after early spring grazing

Following early spring grazing, reduce the 1st cut recommendation by 25 kg/ha. For grazing after cutting, see recommendations in the *Grazing and cutting* section later in the sheep section.

Table 8.8 provides the whole season total nitrogen requirements for cutting and grazing in the *Very poor/Poor* Grass Growth Class. Values in the table are for a moderate SNS situation. **Reduce total fertiliser nitrogen input by 30 kg N/ha for high SNS. Increase total fertiliser nitrogen input by 30 kg N/ha for low SNS (uncommon for dairy production).**

Table 8.8 Sheep: Grass Growth Class *Very poor/Poor*. Whole-season total nitrogen requirement for cut and grazed grass (kg N/ha).

SHEEP Grass Growth Class *Very poor/Poor*	Stocking Rate		Total N requirement	
			Cut	Grazed
	LU/ha	kg/ha	Indicative yield* (t DM/ha)	kg/ha
Moderately to Intensively grazed	1.4	280	7.0	270
	1.2	270	6.9	190
	1.0	260	6.7	140
Extensively to Moderately grazed	0.9	250	6.8	120
	0.7	240	6.5	70
	0.5	230	6.3	30
Extensively grazed	0.2	210	6.0	0
	0.1	200	5.8	0
	0.05	200	5.8	0

- Values in the above table are total annual nitrogen requirement. To obtain your fertiliser requirement, DON'T FORGET to consider nitrogen supply from SNS (page 188), clover (page 208) and applied organic manures (Section 2)
- *Average dry matter yields for cut grass should be achievable in most seasons.
- For estimation of LU for sheep systems, use an average of 0.17 LU per ewe plus lambs. Stocking rates >2.0 increase the risk of high worm burden with consequences for animal productivity.
- For stocking rates that are intermediate between values in the table, assume a proportional difference in nitrogen requirement between two values.

Section 8: Grass

The total nitrogen requirement should be split into 3-6 applications during the growing season if > 150 kg N/ha. Three applications would be appropriate only where the total requirement is less than or equal to 150 kg N/ha. Applications need not be of equal size and advantage can be taken of the relatively high grass growth rate in late spring. For example, the recommended split of N application could be; for **first cut**; 40% (could be split further, Feb-March 15%, April 25%), for **second cut**; 35% (could be split further, May 20%, June 15%); for **subsequent cuts**; 25% (could be split further, July 15%, August 10%).

In mild areas where earlier grazing is possible, nitrogen may be applied from early-mid February. In upland areas, apply nitrogen from mid-late March. Typically, nitrogen would be applied around one month before normal turn-out date.

Cutting after early spring grazing

Following early spring grazing, reduce the 1st cut recommendation by 25 kg/ha. For grazing after cutting, see recommendations in the *Grazing and cutting* section below

Grazing and cutting systems

Where cut grass is followed by grazing, farmers should obtain the two total N requirement values (cut and grazed) for their chosen system. For cut grass, the recommended split of N application could be; For first cut; 40% (could be split further, Feb-March 15%, April 25%), For second cut; 35% (could be further split May 20%, June 15%); subsequent cuts 25% (could be further split July 15%, August 10%). For grazed grass the recommended split could be: Feb-March 15%, April 25%, May 20%, June 15%; July 15%, August 10%.

Then, for example in the case of a 1 cut system followed by grazing; follow the guidance on proportional total N split for 1st cut, i.e. 40% of the total value. Then revert to 20% proportional split of grazed grass total value (starting May) and follow grazing split accordingly.

Similarly, for a 2 cut system followed by grazing; follow the guidance on the proportional total N split for 1st and 2nd cuts, i.e. 40% for 1st cut and 35% for 2nd cut. Then revert to the 15% proportion split of the grazed grass total value (starting July) and follow grazing split accordingly.

Section 8: Grass

Grazing of Grass/Clover Swards – Nitrogen

Generally little fertiliser nitrogen is needed on swards with an appreciable clover content. On average, a good grass/clover sward will give annual dry matter yields equivalent to that produced from about 180 kg N/ha applied to a pure grass sward. However newly sown grass clover leys may yield even more nitrogen depending on the percentage of clover. It is often difficult to decide how much nitrogen will be supplied from grass/clover as the clover content can be very changeable from year to year and within a given season. The following photographs indicate how to estimate clover content and assess nitrogen supply.

Assessing nitrogen supply from clover

Percentage ground cover from clover	Potential nitrogen supply
 20-30% Cover	180 kg/ha
 ~40% Cover	240 kg/ha
 50-60% Cover	300 kg/ha

Section 8: Grass

These figures should be used as rough guides only as full clover development does not normally take place until late spring onwards. For this reason it is only possible to estimate nitrogen supply from clover retrospectively.

Applications of fertiliser nitrogen to purpose-sown grass/clover swards should be made with caution as any form of mineral nitrogen inhibits nitrogen fixation from the atmosphere by rhizobia in the clover nodules. However some nitrogen may need to be applied to grass/clover swards to encourage early spring or late autumn growth.

- Apply up to 50 kg N/ha in mid February to early March if early grass growth is required
- Apply up to 50 kg N/ha in late August to early September if autumn grass is required

Unnecessary use of nitrogen can significantly reduce the percentage of nitrogen-fixing clover in the sward. Clover is particularly sensitive to nitrogen application during establishment. No nitrogen should be used during this period. However, some modern varieties of clover respond to fertiliser nitrogen and are grown in high yields as a source of protein. At high levels of fertiliser N, these clovers will fix little extra nitrogen.

Section 8: Grass

Grazed Grass – Phosphate, Potash and Magnesium (Total for the Whole Year)

	P or K Index			
	0	1	2	3
	kg/ha			
Phosphate (P_2O_5)[a]	80	50	20	0
Potash (K_2O)[b]	60	30	0	0

a. Phosphate may be applied in several small applications during the season, though there may be a small response if it is all applied in early spring for the 1st grazing
b. Potash may either be applied in one application before the third grazing, or in several small applications during the season. At Index 0, apply 30 kg K_2O/ha for the first grazing. Where there is a known risk of hypomagnesaemia, application of potash in spring should be avoided.

For magnesium recommendations, see page 183.

Newly sown swards

In the first season after autumn or spring sowing, deduct the amounts of phosphate and potash applied to the seedbed.

Don't forget to deduct nutrients applied as organic manures – see Section 2

Cutting of Grass/Clover Swards, Red Clover or Lucerne – Nitrogen

Do not apply any nitrogen if a silage crop is taken from a grass/clover sward where the clover content needs to be maintained. Do not apply any fertiliser nitrogen for red clover or lucerne conservation.

Apply phosphate, potash and magnesium as recommended for pure grass swards.

Section 8: Grass

Grass Silage – Phosphate, Potash, Magnesium and Sulphur

The amounts of phosphate and potash are appropriate to the fresh weight yields shown. The yields are based on wilted silage at 25% dry matter content as carted into a silage clamp. Where yields are likely to be greater or smaller, phosphate and potash applications should be adjusted accordingly. Appendix 5 gives typical values of the phosphate and potash content in crop material per tonne of yield.

Different potash recommendations are given for the lower half (2-) and upper half (2+) of K Index 2. When considering the maintenance recommendations (marked 'M'), full account must be taken of note b) beneath the table.

	P or K Index				
	0	1	2	3	4 and over
	kg/ha				
1st cut (23 t/ha)					
Phosphate $(P_2O_5)^a$	100	70	40M	20	0
Potash $(K_2O)^b$ – previous autumn – spring	60 80	30 80	0 80M (2-) 60 (2+)	0 30	0 0
2nd cut (15 t/ha)					
Phosphate $(P_2O_5)^a$	25	25	25M	0	0
Potash $(K_2O)^b$	120	100	90M (2-) 60 (2+)	40	0
3rd cut (9 t/ha)					
Phosphate $(P_2O_5)^a$	15	15	15M	0	0
Potash $(K_2O)^b$	80	80	80M (2-) 40 (2+)	20	0
4th cut (7 t/ha)					
Phosphate $(P_2O_5)^a$	10	10	10M	0	0
Potash $(K_2O)^b$	70	70	70M (2-) 40 (2+)	20	0

a. At soil P Index 2 or over, the whole of the total phosphate requirement may be applied in the spring. At P Index 0 and 1, the phosphate recommendation for the 3rd and 4th cuts may be added to the 2nd cut recommendation and applied in one dressing.

b. To minimise luxury uptake of potash, no more than 80-90 kg K_2O/ha should be applied in the spring for the 1st cut. The balance of the recommended rate should be applied in the previous autumn.

Section 8: Grass

At soil K Indices 2+ or below, extra potash is needed after cutting as follows:

- In one or two cut systems apply an extra 60 kg K_2O/ha following the last cut or by the autumn. Where grazing follows cutting, this may be applied as an extra 30 kg K_2O/ha per grazing for up to two grazings.
- In three cut systems, apply an extra 30 kg K_2O/ha after cutting.
- In four cut systems, no extra potash is needed.

For magnesium recommendations, see page 183.

Soil and herbage analysis to assess potash use

Since large quantities of potash can be removed in intensive silage systems, soil analysis **at least every 4 years** is of particular importance to monitor changes in soil K Indices and adjust potash fertiliser use accordingly. Because soil K levels can fluctuate widely during the growing season, it is important that soil sampling is carried out during the winter or early spring period, leaving the longest possible time interval between sampling and the last potash fertiliser or manure application.

Herbage analysis can also be useful to assess the adequacy of recent potash applications and as a basis for adjusting potash use for future cuts. Uncontaminated samples of herbage should be taken just before cutting. Potash deficiency is indicated if the herbage potassium concentration is below 2% K (in dry matter) or the N:K ratio of the herbage is above 1 to 1.3.

Sulphur

Sulphur deficiency is common in grassland, especially at second and later cuts in multi-cut silage systems using high rates of nitrogen. Deficiency is possible on all mineral soil types though is likely to be particularly severe on sandy and shallow soils in areas of low atmospheric deposition. Deficiency at first cut is less common but can occur on light sand and shallow soils. The map on page 43 shows areas likely to be deficient in sulphur. Grass on organic or peaty soils is not likely to show sulphur deficiency.

The symptoms of sulphur deficiency are a general paling of growth similar to nitrogen deficiency. Analysis of uncontaminated herbage sampled just before cutting is a useful indicator of deficiency. The information can be used to assess the need for sulphur for future cuts. The critical level is 0.25% total sulphur or an N:S ratio greater than 13:1.

Where sulphur deficiency is indicated, apply 40 kg SO_3/ha as a sulphate containing fertiliser applied at the start of growth before each cut.

Newly sown swards

In the first season after autumn or spring sowing, deduct the amounts of phosphate and potash applied to the seedbed.

Don't forget to deduct nutrients applied as organic manures – see Section 2

Section 8: Grass

Hay – Nitrogen

	Soil Nitrogen Supply		
	Low	Moderate	High
	kg/ha		
Each hay cut	100	70	40

Hay – Phosphate, Potash and Magnesium

Different potash recommendations are given for the lower half (2-) and upper half (2+) of K Index 2.

	P or K Index				
	0	1	2	3	Over 3
	kg/ha				
Phosphate (P$_2$O$_5$)	80	55	30M	0	0
Potash (K$_2$O)	140	115	90M (2-) 65 (2+)	20	0

For magnesium recommendations, see page 183.

Newly sown swards

In the first season after autumn or spring sowing, deduct the amounts of phosphate and potash applied to the seedbed.

Don't forget to deduct nutrients applied as organic manures – see Section 2.

June 2010

Section 9: Other useful sources of information

Fertiliser and lime use

Facts
Contact FACTS at the BASIS Office (tel. 01335 3430945) www.basis-reg.com and www.factsinfo.org.uk.

Nutrient management
Information and downloadable documents at www.nutrientmanagement.org

- *Tried and Tested Nutrient Management Plan* – a paper-based plan with optional electronic recording sheets, created by Industry (AIC, CLA, FWAG, LEAF and NFU) and recognised by Defra, EA, levy bodies etc.
- *Part of the Solution – climate change, agriculture and land management* (AIC, CLA, NFU).

Agricultural Industries Confederation
Confederation House, East of England Showground, Peterborough, PE2 6XE (tel 01733 385230) www.agindustries.org.uk

- *Fertiliser Spreaders – Choosing, Maintaining & Using.*
- *Code of Practice for the prevention of water pollution from the storage and handling of solid fertilisers.*
- *Code of Practice for the prevention of water pollution from the storage and handling of fluid fertilisers.*
- *British Survey of Fertiliser Practice* (annual priced publication).

The Agricultural Lime Association
tel 01733 385240 www.aglime.org.uk

Agricultural lime – the natural solution.

Potash Development Association (PDA)
PO Box 697, York YO32 5WP www.pda.org.uk

- *Phosphate and potash removal by crops* and other technical literature on the use of phosphate and potash.

International Fertiliser Society (IFS)
PO Box 4, York YO32 5YS www.fertiliser-society.org

- *Lime, liming and the management of soil acidity* [ISBN 978-0-85310-044-7] and other technical literature on fertiliser production and use.

Section 9: Other useful sources of information

British Standards Institute

BSI British Standards, 389 Chiswick High Road, London, W4 4AL (tel. 020 8996 9001)

- *Specification for topsoil and requirements for use.* British Standard BS 3882: 2007, ISBN: 978 0 580 61009. Publication Date: 30th Nov 2007.

- *The Analysis of Agricultural Materials. MAFF RB427.* This book is out of print but copies are available through libraries.

BCPC Publications Sales

tel: +44 0 1420 593 200

- *Spreading Fertilisers and Applying Slug Pellets.*

Environment Agency

www.environment-agency.gov.uk

- think**soils.**

FAO

FAO Sales and Marketing Group, Viale delle Terme di Caracalla, 00153, Rome, Italy. www.fao.org

- Efficiency of soil and fertiliser phosphorus use. *FAO Fertilizer and Plant Nutrition Bulletin 18* (2008). Syers J K, Johnston A E and Curtin D.

Soil Science

- Hislop J and Cooke I J (1968). Anion exchange resin as a means of assessing soil phosphate status: a laboratory technique. *Soil Science*, volume 105, 8-11.

Advances in Agronomy

- Johnston A E, Poulton P R and Coleman K. (2009). Soil organic matter: its importance in sustainable agriculture and carbon dioxide fluxes. *Advances in Agronomy*, volume 101.

National Soil Resources Institute

Building 63, Cranfield University, Beds, (tel. 01234 754086).

- Soil survey publications.

Section 9: Other useful sources of information

Organic manure use

Defra

The following are available free from Defra Publications, Admail 6000, London SW1A 2XX (tel. 08459 556000).

- *Opportunities for saving MONEY by reducing WASTE on your farm (2000), PB4819*
- *Ammonia in the UK (2002) PB6865*

ADAS

Gleadthorpe, Meden Vale, Mansfield, Nottingham, NG20 9PF (tel. 01623 844331)

- MANNER-*NPK* computer programme www.adas.co.uk/manner.
- *Controlling soil erosion – A manual for the assessment and management of agricultural land at risk of water erosion in lowland England (1999), PB4093.*
- *Managing Livestock Manures Booklet 1 (Revised 2001) – Making better use of livestock manures on arable land.*
- *Managing Livestock Manures Booklet 2 (Revised 2001) – Making better use of livestock manures on grassland.*
- *Managing Livestock Manures Booklet 3 (2001) – Spreading systems for slurries and solid manures.*
- *Managing Livestock Manures Booklet 4 (2001) – Managing manure on organic farms.*

Code of practice and regulations

Defra

Available free from Defra Publications, Admail 6000, London SW1A 2XX (tel. 0171 238 5665):

- *Code of Practice for Agricultural Use of Sewage Sludge (1996).*
- *Guidance for Farmers in Nitrate Vulnerable Zones (2008) PB12736 a - i.*

Statutory Instruments

The Fertilisers Regulations 1991 (SI 2197); The Fertilisers (Amendment) Regulations 1995, No 16; The Fertilisers (Amendment) Regulations 1997, No 1543; The Fertilisers (Amendment) Regulations 1998, No 2024; The Fertilisers (Sampling and Analysis) Regulations 1996 (SI 1342); The Ammonium Nitrate Materials (High Nitrogen Content) Safety Regulations 2003 (SI 1082); The EC Fertilisers (England and Wales) Regulations 2006 (SI 2486). www.legislation.hmso.gov.uk/stat.htm

Section 9: Other useful sources of information

EA

Integrated Pollution Prevention and Control (IPPC): Intensive Farming. How to comply – Guidelines for Intensive Pig and Poultry Farmers, April 2006. Available from the Environment Agency, Riotlouse. Waterside Drive, Aztec West, Almendsbury, Bristol BS32 4UD. www.environment-agency.gov.uk.

ADAS/Water UK

The following are available from Water UK, 1 Queen Anne's Gate, London, SW1H 9BT (Tel: 020 7344 1844) or ADAS Gleadthorpe Research Centre, Meden Vale, Mansfield, Nottingham, NG20 9PF (tel. 01623 844331).

The Safe Sludge Matrix – Guidelines for the Application of Sewage Sludge to Agricultural Land, 3rd edition April 2001 www.adas.co.uk

TSO (The Stationery Office)

The following are available from TSO (tel. 0870 6005522, www.tso.co.uk).

Protecting Our Water, Soil and Air: A Code of Good Agricultural Practice for farmers, growers and land managers (Defra, 2009).

The Sludge (Use in Agriculture) Regulations 1989 (SI 1263); The Sludge (Use in Agriculture) (Amendment) regulations 1990 (SI 880).

The Pollution Prevention and Control (England and Wales) Regulations 2000 (SI 800).

Appendix 1: Description of Soil Types

Soil Category	Description of Soil Types Within Category
Light sand soils	Soils which are sand, loamy sand or sandy loam to 40 cm depth and are sand or loamy sand between 40 and 80 cm, or over sandstone rock
Shallow soils	Soils over impermeable subsoils and those where the parent rock (chalk, limestone or other rock) is within 40 cm of the soil surface. Sandy soils developed over sandstone rock should be regarded as light sand soils.
Medium soils	Mostly medium-textured mineral soils that do not fall into any other soil category. This includes sandy loams over clay, deep loams, and silty or clayey topsoils that have sandy or loamy subsoils.
Deep clayey soils	Soils with predominantly sandy clay loam, silty clay loam, clay loam, sandy clay, silty clay or clay topsoil overlying clay subsoil to more than 40cm depth. Deep clayey soils normally need artificial field drainage.
Deep silty soils	Soils of sandy silt loam, silt loam or silty clay loam textures to 100 cm depth or more. Silt soils formed on marine alluvium, warp soils (river alluvium) and brickearth soils are in this category. Silty clays of low fertility should be regarded as other mineral soils.
Organic soils	Soils that are predominantly mineral but with between 10 and 20% organic matter to depth. These can be distinguished by darker colouring that stains the fingers black or grey.
Peat soils	Soils that contain more than 20% organic matter derived from sedge or similar peat material

Appendix 1: Description of Soil Types

Assessment of Soil Texture

Accurate measurement of soil texture requires laboratory analysis, but for practical purposes texture can be assessed by hand using the following method:

Take about a dessert spoonful of soil. If dry, wet up gradually, kneading thoroughly between finger and thumb until soil crumbs are broken down. Enough moisture is needed to hold the soil together and to show its maximum stickness. Follow the paths in the diagram to get the texture class.

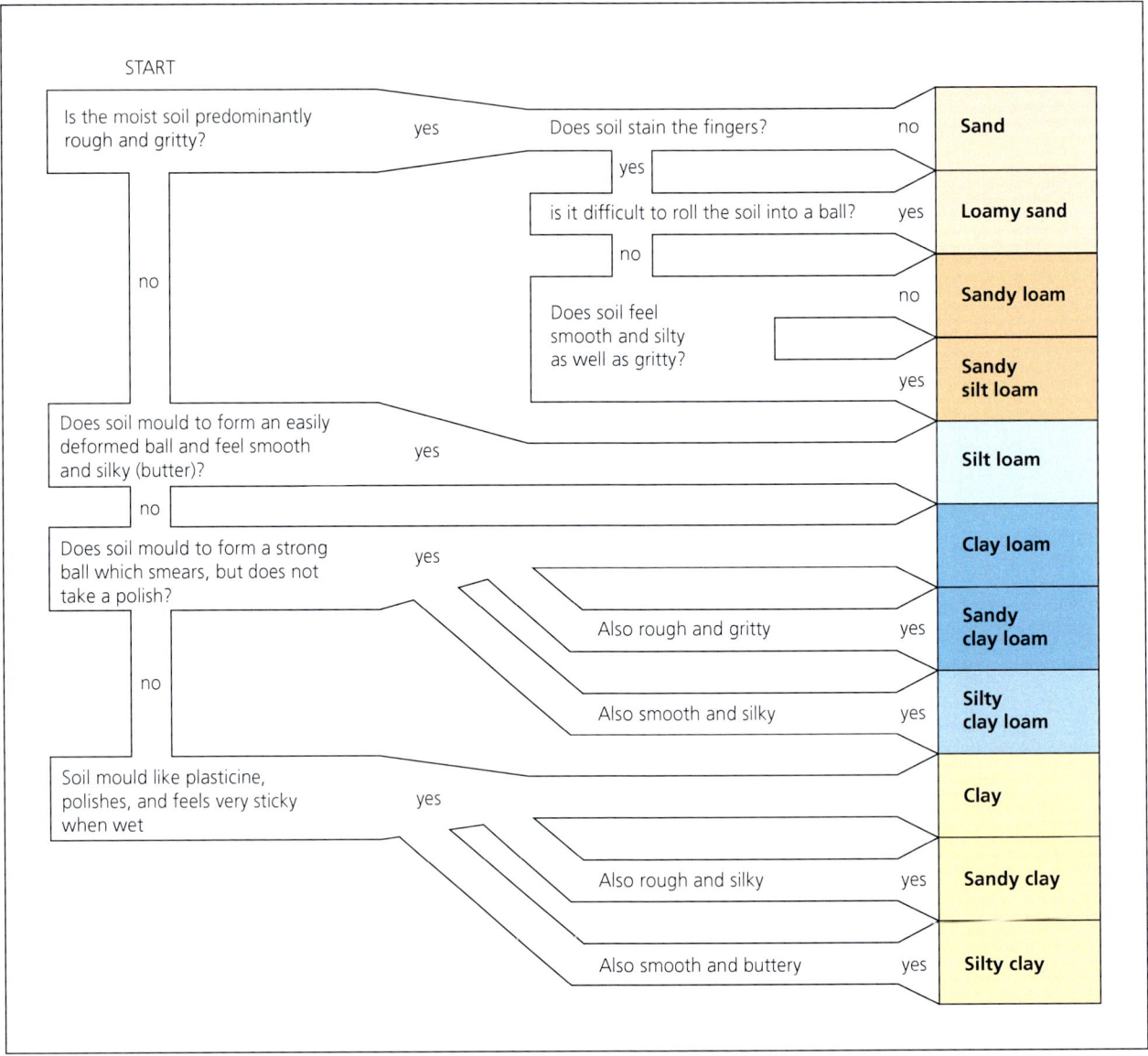

A texture triangular diagram, defining the particle size distribution for each named texture class, is given in Appendix D of *Controlling Soil Erosion (MAFF PB4093)*.

Appendix 2: Sampling for Soil Mineral Nitrogen (Smn); Estimation of Crop Nitrogen Content and Mineralisable Nitrogen

Direct measurement or estimation of the components of Soil Nitrogen Supply (SNS) provides a more reliable basis for nitrogen decisions in a number of situations. Section 3 describes the importance of SNS in more detail.

It is important not to confuse SNS (Soil Nitrogen Supply) and SMN (Soil Mineral Nitrogen). Although SMN is usually the most important component of SNS and the one that is commonly measured by laboratory analysis, calculation of the SNS needs measurement or estimation of **all** components of the SNS using the guidelines below.

> *Soil Nitrogen Supply (SNS) = SMN + estimate of total crop nitrogen content + estimate of net mineralisable nitrogen*
>
> where:
>
> - Soil Mineral Nitrogen (kg N/ha) is the nitrate-N plus ammonium-N content of the soil within the potential rooting depth of the crop.
> - Total crop nitrogen (kg N/ha) is the total content of nitrogen in the crop (if present) when sampling for SMN is carried out.
> - Net mineralisable nitrogen (kg N/ha) is the estimated amount of nitrogen made available for crop uptake from mineralisation of soil organic matter and crop debris during the growing season.

Sampling and analysis for SMN

SNS indices in this *Manual* are based on soil sampling to 90cm for SMN The results of analysis for SMN will be meaningful only if the sample is taken correctly and analysed using recognised procedures. In general, laboratory analysis complying with the guidelines for good laboratory practice is accurate. The main source of variability/uncertainty occurs when taking the soil samples. The results of analysis on badly taken samples, especially when an inadequate number of soil cores is taken, may be misleading and can lead to expensive mistakes if wrong fertiliser decisions are made.

Other soil sampling systems are available and if they are used only their own associated recommendation systems should be used.

Targeting fields for sampling

Since sampling and analysis is costly, it is not realistic to measure SMN in every field in every year. Measurement will usually be most worthwhile in fields where large and unpredictable soil nitrogen residues are expected. Typical examples are:

- Where organic manures have been regularly used.
- Following crops where large quantities of nitrogen-rich, leafy crop debris have been left.

Appendix 2: Sampling for Soil Mineral Nitrogen (Smn); Estimation of Crop Nitrogen Content and Mineralisable Nitrogen

- Where long leys or permanent pasture have been recently ploughed out (but not in the first year after ploughing out).
- Where there have been problems such as regular lodging of cereals or previous crop failure (for example due to drought or disease).
- Fields where there is significant variation in soil texture and/or large amounts of crop residues are incorporated.

Analysis for SMN is **not** recommended on peaty soils or in the first season after ploughing out long leys or permanent pasture. In these situations, nitrogen released by mineralisation of soil organic matter is a large component of the SNS and is difficult to predict accurately.

The amount of SMN is likely to be larger on nitrate 'retentive' medium, clay or silty soils than on nitrate 'leaky' sandy or shallow soils, in which SMN levels are usually low and more predictable. Sampling to determine SNS Index is not currently recommended in grassland systems or for established fruit, vines or hops.

Taking the soil sample

Time of sampling

In most situations, sampling in early late winter or early spring before nitrogen fertiliser is applied, is preferable especially in high rainfall areas or on shallow or light sand soils. On soils less prone to leaching, sampling earlier in winter is possible. Avoid sampling within 2-3 months after application of nitrogen fertiliser or organic manures.

Depth of sampling

Sampling to the full 90 cm is recommended when sampling from January onwards. Sampling to 60 cm is satisfactory when sampling in the autumn. If sampling to 60 cm and deep-rooted crops are to be grown, an estimate of the mineral nitrogen in the 60-90 cm layer must be made. In the absence of any better estimate, uniform distribution of nitrogen in the profile should be assumed.

For shallow rooted crops (e.g. some vegetables), sampling to 30 cm or to rooting depth is all that is needed. However, to identify the correct SNS Index will require an estimate of mineral nitrogen to 90cm.

Method of sampling

Samples must be taken to be representative of the area sampled. A minimum of 15-20 soil cores should be taken per field (based on a 10 ha field) and bulked to form a representative sample. Thorough mixing of the bulked sample is vital prior to sub-sampling for sending to the laboratory. Areas of land known to differ in some important respects (e.g. soil type, previous cropping, application of manures or nitrogen fertiliser) should be sampled separately. In large fields (more than 10 ha), especially where the soil type is not uniform, more than one sample should be taken. This can be done by dividing the field into smaller blocks from each of which 10-15 soil cores are taken.

Appendix 2: Sampling for Soil Mineral Nitrogen (Smn); Estimation of Crop Nitrogen Content and Mineralisable Nitrogen

It is important to avoid cross-contamination of samples from different depths. Using a mechanised 1 metre long gouge auger (2.5 cm diameter) is a satisfactory and efficient method but care must be taken to avoid excessive soil compaction and contamination between soil layers. If each depth layer is to be sampled individually by hand, a series of screw or gouge augers should be used where the auger diameter becomes progressively narrower as the sampling depth increases.

Transport to the laboratory

Soil mineral nitrogen concentrations can change during storage of samples. After sampling, soils should **not** be frozen but be kept refrigerated at less than 5°C and remain cooled while transported as quickly as possible to the laboratory. Samples should remain cooled until analysis which should be carried out as soon as possible after sampling.

Analysis in the laboratory

Samples should be analysed for nitrate-N and ammonium-N. Potassium chloride is a suitable extractant. The analytical results (mg N/kg) will need to be converted into kg/ha of mineral nitrogen in each soil layer and then summed to give a value for the whole soil profile. Ideally, soil weights should be calculated from bulk density measurements. However, this is rarely practical to 90cm depth so standard values must be used. Normally, the laboratory undertaking analysis for SMN will use standard values.

Estimating the crop nitrogen content

It is important to estimate the nitrogen content of any crop that is present at the time of soil sampling. This is often a small though important component of the SNS. In cereal crops, this can be assessed according to the number of shoots present (main shoots and tillers), as follows:

Shoot number/m^2	Crop nitrogen content (kg N/ha)
500	5-15
1000	15-30
1500	25-50

Use the lower figure when assessing crops in late autumn and the higher figure for crops in early spring.

In oilseed rape, the nitrogen content of an average density crop can be assessed by measuring the average crop height.

Crop height (cm)	Crop nitrogen content (kg N/ha)
10	35-45
15	55-65
20	75-85

Appendix 2: Sampling for Soil Mineral Nitrogen (Smn); Estimation of Crop Nitrogen Content and Mineralisable Nitrogen

Estimating the nitrogen that will be released from mineralisation of organic matter

Nitrogen mineralised from soil organic matter and crop debris after soil sampling is a potentially important source of nitrogen for crop uptake, though in mineral soils of low to average organic matter content, the amount of mineralisable nitrogen will be small and for practical purposes, can be ignored.

Research has not yet identified a preferred laboratory method that is suitable for routine use. Measurement of the topsoil organic matter content, anaerobic incubation techniques or computer modelling can give useful indications. As a guide, a soil that has a topsoil organic matter content of 10% may release 60-90 kg N/ha more potentially available nitrogen than an equivalent soil with 3% organic matter content, or 150-200 kg N/ha where the topsoil organic matter content is 20%. This nitrogen becomes available gradually through the year. In contrast, the nitrogen in crop debris is released rapidly following incorporation into the soil.

Appendix 3: Sampling for Soil pH, P, K, Mg and Na

The results of a soil analysis will only be meaningful if the sample is taken in the field in the correct way and analysed in the laboratory using recognised analytical procedures. The recommendations in this *Manual* are based on the use of specific soil analysis procedures and should not be used where other analytical methods have been used. In general, laboratory analysis is very accurate and the main source of error is when taking the soil sample in the field. The results of soil analysis from badly taken samples may be misleading and can lead to expensive mistakes if wrong fertiliser decisions are made.

The following sampling procedures should be used when sampling arable, vegetable or grassland. Recommendations for sampling fruit, vines and hops are given in Section 6:

When to sample

In most systems, soil pH and nutrient levels other than inorganic nitrogen change slowly, so it is not necessary to resample and analyse every year. In general, sampling every fourth year is satisfactory as a basis for fertiliser recommendations, but pH may need more frequent monitoring.

Sampling must be carried out at a time when the soil nutrient status is in a settled state. To allow meaningful comparison between analysis results from different four-year cycles, a sampling strategy should be developed, so that samples are taken at the same point in the rotation and with respect to recent fertiliser or manure applications and soil cultivations.

- Leave as long as possible between the last fertiliser or manure application. If possible, sample **after** the last fertiliser or manure application has been cultivated into the soil.
- Do **not** sample within six months of a lime or fertiliser application (except nitrogen).
- Sample at the same point in the rotation and well before growing a sensitive crop (e.g. sugar beet).
- Avoid sampling when the soil is very dry.

Taking a representative sample

The soil sample must be representative of the area sampled. Areas of land known to differ in some important respects (e.g. soil type, previous cropping, applications of manure, fertiliser or lime) should be sampled separately. Small areas known to differ from the majority of a field should be excluded from the sample.

A sample of 25 individual sub-samples (cores) will be adequate for a uniform area. The sub-sampling points must be selected systematically, with an even distribution over the whole area. This may be achieved by following the pattern of a letter 'W' and taking sub-samples at regular intervals. Do not take samples in headlands, or in the immediate vicinity of hedges, trees or other unusual features.

Appendix 3: Sampling for Soil pH, K, Mg and Na

Sampling depth

Uniformity of sampling depth is particularly important where crops are established without ploughing or in established grassland. Where there is little or no mixing of the topsoil, nutrients from fertiliser and manures tend to remain in the surface few centimetres of the soil.

There is a standard depth for sampling depending on the crop rotation.

Arable and field vegetables	Sample to 15 cm depth
Long term grassland	Sample to 7.5 cm depth

Sampling equipment

A gouge corer or screw auger may be used when sampling in arable or vegetable systems, or for fruit, vines and hops.

In grassland systems or where the soil is not cultivated, only use a gouge or pot corer which can take an even core of soil throughout the sampling depth. This is not possible using a screw auger which should **not** be used in these situations.

Acid soils

On soils where acidity is known to occur, more frequent testing may be needed than the four-year cycle used for phosphate, potash and magnesium. Since acidity can occur in patches, spot testing with soil indicator across the field is often useful. Soil indicator can also be useful on soils which contain fragments of free lime, since these can give a misleadingly high pH when analysed following grinding in the laboratory.

Appendix 4: Classification of Soil P, K and Mg Analysis Results into Indices

Index	Phosphorus (P)		Potassium (K)	Magnesium (Mg)
	Olsen's P	Resin P	Ammonium nitrate extract	
	mg/litre			
0	0-9	0-19	0-60	0-25
1	10-15	20-30	61-120	26-50
2	16-25	31-49	121-180 (2-) 181-240 (2+)	51-100
3	26-45	50-85	241-400	101-175
4	46-70	86-132	401-600	176-250
5	71-100	>132	601-900	251-350
6	101-140		901-1500	351-600
7	141-200		1501-2400	601-1000
8	201-280		2401-3600	1001-1500
9	Over 280		over 3600	over 1500

This classification should only be used where the methods of laboratory analysis used follow the procedures described in *Specification for topsoil and requirements for use*. British Standard BS 3882: 2007 or *The Analysis of Agricultural Materials (MAFF RB427)* or, for resin P, the method described in Hislop and Cooke (1968). *Anion exchange resin as a means of assessing soil phosphate status: a laboratory technique* (details of all of these in Section 9).

Appendix 5: Phosphate and Potash in Crop Material

		Phosphate (P$_2$O$_5$)	Potash (K$_2$O)
		kg/t of fresh material	
Cereals	Grain only (all cereals)	7.8	5.6
	Grain and straw		
	- winter wheat/barley *	8.4	10.4
	- spring wheat/barley *	8.6	11.8
	- winter/spring oats *	8.8	17.3
Oilseed rape	Seed only	14.0	11.0
	Seed and straw *	15.1	17.5
Peas	Dried	8.8	10.0
	Vining	1.7	3.2
Field beans		11.0	12.0
Straw**	Winter wheat, winter barley	1.2	9.5
	Spring wheat, spring barley	1.5	12.5
	Oilseed rape	2.2	13.0
	Beans	2.5	16.0
	Peas	3.9	16.0
Potatoes		1.0	5.8
Sugar beet	Roots only	0.8	1.7
	Roots and tops	1.9	7.5
Grass	Fresh grass (15-20% DM)	1.4	4.8
	Silage (25% DM)	1.7	6.0
	Silage (30% DM)	2.1	7.2
	Hay (86% DM)	5.9	18.0
Kale		1.2	5.0
Maize	Silage (30% DM)	1.4	4.4
Swedes	Roots only	0.7	2.4
Broad beans		1.6	3.6
French beans		1.0	2.4
Beetroot		1.0	4.5
Cabbage		0.9	3.6
Carrots		0.7	3.0
Cauliflowers		1.4	4.8
Onions	Bulbs only	0.7	1.8
Sprouts	Buttons	2.6	6.3
	Stems	2.1	7.2
Bulbs		2.4	6.3

* Offtake values are per tonne of grain or seed removed but include nutrients in straw when this also is removed without weighing (see first example below).

** These values to be used only when straw weight is known. Potash content of straw can vary substantially – higher than average rainfall between crop maturity and baling straw will reduce straw potash content. There is less information on phosphate and potash contents in the non-cereals so the values above should be treated as guides only.

Appendix 5: Phosphate and Potash in Crop Material

> **Example**
>
> *Winter wheat yields 10 t/ha of grain. The straw is baled and removed from the field.*
>
> Phosphate offtake = 10 x 8.4 = 84 kg P_2O_5/ha
> Potash offtake = 10 x 10.4 = 104 kg K_2O /ha

> **Example**
>
> *Spring barley straw is baled and 5 t/ha removed from the field.*
>
> Phosphate offtake = 5 x 1.5 = 7.5 kg P_2O_5/ha
> Potash offtake = 5 x 12.5 = 62 kg K_2O /ha

Appendix 6: Sampling organic manures for analysis

The nutrient content of slurry can vary considerably within a store due to settlement and crusting. In particular, pig slurry can 'settle out' in storage, with a higher dry matter layer being at the base of the store and a lower dry matter layer occupying the mid/upper part of the store, which during store emptying can markedly affect slurry dry matter and associated nutrient contents. Similarly, the composition of solid manure in a heap can vary depending on the amount of bedding and losses of nutrients during storage. If stored materials are to be analysed either in a laboratory or using a rapid on-farm method (e.g. using Agros or Quantofix slurry-N meters), it is important that the sample taken represents an 'average' of what is found in the heap or store.

Taking representative samples

Sampling stores or heaps

It is important, where practical and safe, to take several sub-samples. Take these from a range of positions within the store or heap, bulk them together and then take a representative sub-sample. Send this to the laboratory for analysis or test it on-farm with a slurry-N meter or slurry hydrometer.

Slurries

Take at least five sub-samples of 2-litres, pour into a large container, stir thoroughly, and pour a 2-litre sub-sample immediately into a smaller clean container. This is the sample for analysis.

Above ground stores

Ideally, slurry should be fully agitated and sub-samples taken from the reception pit. If this is not possible, and provided there is safe access from an operator's platform, the five sub-samples can be taken at a range of positions, using a weighted 2 litre container attached to a rope.

Below ground pit

It may be possible to obtain sub-samples at various positions using a weighted container as above.

Earth-banked lagoons

Do not attempt to sample direct from the lagoon unless there is a secure operator's platform which provides safe access. If the slurry has been well agitated, sub-samples can be obtained from the slurry tanker or irrigator. If the tanker is fitted with a suitable valve, it may be possible to take five sub-samples from this stationary tanker at intervals during filling or while field spreading is in progress.

When sampling enclosed slurry stores (pits or tanks) never climb down or lean into the store because of the risk of inhaling toxic gases.

Appendix 6: Sampling organic manures for analysis

Solid Manures

Take at least 10 sub-samples of about 1 kg each as described below, and place on a clean, dry tray or sheet. Break up any lumps and thoroughly mix the sample. Then take a representative sample of around 2 kg for analysis.

Manure heaps

Provided the manure is dry and safe to walk on, identify at least ten locations which appear to be representative of the heap. After clearing away any weathered material with a spade or fork, dig a hole approximately 0.5 metres deep and take a 1 kg sample from each point. Alternatively, take sub-samples from the face of the heap at various stages during spreading.

Weeping wall stores

Do not attempt to take samples before the store is emptied as it is not safe to walk on the surface of the stored material. Sub-samples may be taken from the face of the heap once emptying has commenced.

Sampling during spreading

Trays placed in the field can be used to collect samples from a slurry or solid manure spreading system while the material is being spread. Take care to avoid the possibility of injury from stones and other objects which may be flung out by the spreading mechanism.

Sample containers and analysis

On-farm rapid analysis of slurries should be carried out immediately after sampling, making sure that the sample taken is well mixed. Slurry samples sent to a laboratory should be dispatched in clean, screw-topped 2-litre plastic containers. Leave at least 5 cm of airspace to allow the sample to be shaken in the laboratory. Solid manure samples should be transported in 500 gauge polythene bags. Expel excess air from the bag before sealing.

Clearly label all samples on the outside of the container or bag and dispatch them immediately or within a maximum of seven days of sampling if kept in a refrigerator.

You should wear rubber gloves and protective clothing when collecting samples. Remember to wash hands and forearms thoroughly after taking samples and before eating or drinking.

Appendix 6: Sampling organic manures for analysis

Calculations and interpretation of manure laboratory analysis results

Farmers and advisers are increasingly recognising the importance of allowing for the nutrient content of organic manure applications to land. Provided that the sampling and analysis are done carefully, the results of analysis on specific manure samples will usually provide added accuracy compared to using typical values which are derived from analyses of a large number of manure samples.

Typical values for the nutrient content of different types of manures are contained in Section 2 of this *Manual*. Unfortunately, laboratories differ in the way that the analysis results are expressed and conversion of the results is often needed before they can be used or entered in decision support systems such as MANNER-*NPK* or PLANET. Analysis results are variously reported

- on a dry weight (DW or 100% DM) or fresh weight (FW) basis

- in units of grams or milligrams per kilogram (g/kg, mg/kg), grams per 100 grams (g/100g), percent (%), grams or milligrams per litre (g/l, mg/l), kilograms per tonne (kg/t) or kilograms per cubic metre (kg/m^3)

- as the nutrient element (N, P, K, Mg, S) or the nutrient oxide (P_2O_5, K_2O, MgO, SO_3)

If in doubt about how the results are expressed by the laboratory, as a first step you should contact the laboratory and confirm how they express the results.

Use the following conversions if analysis results need to be converted.

Nutrients

To convert nutrient element to nutrient oxide

$$P \times 2.291 = P_2O_5$$
$$K \times 1.205 = K_2O$$
$$Mg \times 1.658 = MgO$$
$$S \times 2.5 = SO_3$$

Solid manures (DM expressed as %)

To convert mg/kg nutrient in DM to kg/t FW

$$\frac{mg/kg\ nutrient}{1000} \times \frac{\%\ DM}{100} = kg/t\ FW$$

To convert g/kg nutrient in DM to kg/t FW

$$g/kg\ nutrient \times \frac{\%\ DM}{100} = kg/t\ FW$$

To convert g/100g nutrient in DM to kg/t FW

$$g/100g\ nutrient \times \frac{\%\ DM}{10} = kg/t\ FW$$

To convert % nutrient in DM to kg/t FW

$$\%\ nutrient \times \frac{\%\ DM}{10} = kg/t\ FW$$

Appendix 6: Sampling organic manures for analysis

Solid manures (DM expressed as g/kg)

To convert mg/kg nutrient in DM to kg/t FW $\quad\dfrac{\text{mg/kg nutrient}}{1000} \times \dfrac{\text{g/kg DM}}{1000} = \text{kg/t FW}$

To convert g/kg nutrient in DM to kg/t FW $\quad \text{g/kg nutrient} \times \dfrac{\text{g/kg DM}}{1000} = \text{kg/t FW}$

To convert g/100g nutrient in DM to kg/t FW $\quad \text{g/100g nutrient} \times \dfrac{\text{g/kg DM}}{100} = \text{kg/t FW}$

To convert % nutrient in DM to kg/t FW $\quad \text{\% nutrient} \times \dfrac{\text{g/kg DM}}{100} = \text{kg/t FW}$

Liquid Manures

To convert mg/l nutrient to kg/m³ $\quad \dfrac{\text{mg/l nutrient}}{1000} = \text{kg/m}^3$

To convert g/l nutrient to kg/m³ $\quad \text{g/l nutrient} = \text{kg/m}^3$ (no change)

Example

Digested sludge cake with 27% DM, and 4.0% N and 3.0% P in DM.

$4.0\%\text{N} \times \dfrac{27\% \text{ DM}}{10} = 10.8 \text{ kg N/t in FW}$

$3.0\%\text{P} \times \dfrac{27\% \text{ DM}}{10} \times 2.29 = 18.6 \text{ kg P}_2\text{O}_5/\text{t in FW}$

FW = fresh weight.

Appendix 7: Analysis of some fertiliser and liming materials

The materials listed below are used individually and some are used as components of compound or multi-nutrient, fertilisers.

Typical percentage nutrient content (can vary slightly in some materials)

Nitrogen fertilisers

Ammonium nitrate	33.5-34.5% N
Liquid nitrogen solutions	18-30% N (w/w)
Calcium ammonium nitrate (CAN)	26-28% N
Ammonium sulphate	21% N, 60% SO_3
Urea	46% N
Calcium nitrate	15.5% N, 19% CaO

Phosphate fertilisers

Triple superphosphate (TSP)	45-46% P_2O_5
Di-ammonium phosphate (DAP)	18% N, 46% P_2O_5
Mono-ammonium phosphate (MAP)	12% N, 52% P_2O_5
Rock phosphate (eg. Gafsa)	27-33% P_2O_5

Potash, magnesium and sodium fertilisers

Muriate of potash (MOP)	60% K_2O
Sulphate of potash (SOP)	50% K_2O, 45% SO_3
Potassium nitrate	13% N, 45% K_2O
Kainit	11% K_2O, 5% MgO, 26% Na_2O, 10% SO_3
Sylvinite	minimum 16% K_2O, typically 29% Na_2O
Kieserite (magnesium sulphate)	25% MgO, 50% SO_3
Calcined magnesite	typically 80% MgO
Epsom salts (magnesium sulphate)	16% MgO, 33% SO_3
Agricultural salt	50% Na_2O

Sulphur fertilisers

Ammonium sulphate	21% N, 60% SO_3
Epsom salts (magnesium sulphate)	16% MgO, 33% SO_3
Elemental sulphur	typically 200-225% SO_3 (80 – 90% S)
Gypsum (calcium sulphate)	40% SO_3

Appendix 7: Analysis of some fertiliser and liming materials

Other

Ashed poultry manure typically	20% P_2O_5, 10% K_2O
Steelmaking slag typically	1.7-2.5% P_2O_5 + NV 55-58

Liming materials	**Neutralising Value (NV)**
Ground chalk or limestone	50-55
Magnesian limestone	50-55, over 15% MgO
Hydrated lime	c.70
Burnt lime	c.80
Sugar beet lime	22-32 + typically 7-10 kg P_2O_5, 5-7 kg MgO, 3-5 kg SO_3/tonne

The chemical and physical forms of nutrient sources, as well as growing conditions, can influence the effectiveness of fertilisers. A FACTS Qualified Adviser can give advice on appropriate forms for different soil and crop conditions.

The reactivity, or fineness of grinding, of liming materials determines their speed of action. However, the amount of lime needed is determined mainly by its neutralising value.

Appendix 8: Conversion tables

Metric to Imperial

1 tonnes/ha	0.4 tons/acre
100 kg/ha	80 units/acre
1 kg/tonne	2 units/ton
10 cm	4 inches
1 m^3	220 gallons
1 m^3/ha	90 gallons/acre
1 kg/m^3	9 units/1000 gallons
1 kg	2 units

Note: a 'unit' is 1% of 1 hundredweight, or 1.12 lbs.

Imperial to metric

1 ton/acre	2.5 tonnes/ha
100 units/acre	125 kg/ha
1 unit/ton	0.5 kg/tonne
1 inch	2.5 cm
1000 gallons	4.5 m^3
1000 gallons/acre	11 m^3/ha
1 unit/1000 gallons	
1 unit	0.5 kg

Element to Oxide

P to P_2O_5	Multiply by 2.291
K to K_2O	Multiply by 1.205
Mg to MgO	Multiply by 1.658
S to SO_3	Multiply by 2.5
Na to Na_2O	Multiply by 1.348
Na to salt	Multiply by 2.542

Oxide to element

P_2O_5 to P	Multiply by 0.436
K_2O to K	Multiply by 0.830
MgO to Mg	Multiply by 0.603
SO_3 to S	Multiply by 0.4
Na_2O to Na	Multiply by 0.742
Salt to Na	Multiply by 0.393

Fluid fertiliser

kg/tonne (w/w basis) to kg/m^3 (w/v basis)	Multiply by specific gravity

Appendix 9: Calculation of the crop nitrogen requirement (CRN) for field vegetable crops

Where sufficient data are available the nitrogen fertiliser recommendations are based on a three-step framework,

- **Size of the crop** – the size, frame, or weight of the crop needed to provide optimal economic production.
- **Nitrogen uptake** – the optimum nitrogen uptake associated with a crop of that size.
- **Supply of Nitrogen** – based on the nitrogen supply from the soil within rooting depth including the potential nitrogen mineralised from soil organic matter.

1) SIZE OF CROP (Expressed as t/ha Dry wt Yield)

Dry wt Yield = *(Fresh wt yield * dm%/100)/DW-HI*

Fresh wt Yield (t/ha) – i.e. the yield of marketable produce removed or expected to be removed from the field in commercial practice. *Data based on field experiments or expert opinion for well grown crops.*

dm % – dry matter % of marketable produce for optimally fertilised crops.

DW-HI – proportion of the whole crop that is taken for market expressed on a dry wt basis. – in onions 81% (0.81) of the green crop is bulb. For sprouts only 26% (0.26) of the crop grown is produced as sprouts.

2) N UPTAKE IN OPTIMALLY GROWN CROP (kg/ha)

Total N Uptake (kg/ha) = *Dry wt Yield t/ha x N% x10*

N% – This is the estimated nitrogen concentration of an optimally fertilised whole plant at harvest. Generally %N declines as yields increase. For small changes in yield if yield increases by 10% it could be assumed that nitrogen uptake increases by the same amount. For bigger changes the increases in nitrogen uptake will be less than the increase in yield and the formulae below should be used.

$$\%N_{Crit} = a(1+be^{-0.26W})$$

Where a and b are parameters controlling the shape of the curve. W = total dry matter yield t/ha.

The data have been derived from experiments – similar relationships have been used to define critical N concentrations for EU-Rotate N and WELL_N models.

3) SUPPLY OF NITROGEN AND CALCULATION OF CROP NITROGEN REQUIREMENT (CNR)

CRN = (NUptake – MineralisedN + SoilMinN x Rdepth/90) FertRec

Mineralised N – amount of nitrogen released from soil organic matter by mineralisation during the cropping period. This is normally between planting and harvesting dates but for some crops such as lettuce and onions where early nitrogen supply is important a shorter period has been chosen see table below.

Appendix 9: Calculation of the crop nitrogen requirement (CRN) for field vegetable crops

The calculations are based on mineralisation within the WELL_N model (i.e 0.7 kg/ha @ 15.9 Degree C scaled for Wellesbourne temperature). When temperature is less than 4 Degrees mineralisation is assumed to be negligible.

RDepth – Rooting Depth, generally related to the Dry matter yield of the crop. Roots 10cm deep per tonne of dry matter yield – exceptions such as onions and leeks. Maximum rooting depth assumed not to be more than 90 cm. There will be circumstances on fertile soils where rooting is much deeper so potential nitrogen supply is larger than assumed in these tables.

SoilMinN – Soil mineral nitrogen to 90 cm depth.

Fert Rec – Fertiliser recovery assumed to be 60%.

Appendix 10: Information for derivation of crop nitrogen requirement of field vegetable crops

Crop	Fresh Market Yield t/ha	% Dry matter Marketable	Dry wt harvest Index	Total dry matter t/ha	Relation N% and dry matter yield 'a'	Relation N% and dry matter yield 'b'	% N	Tot N uptake kg/ha	Mineralised kg/ha	Period dates	Root depth cm	Recovery Fert %
Brussels sprouts	20.3	17.0	0.26	13.3	2.50	3.50	2.8	368	121	20/5-17/12	90	60
White Cabbage Storage	110.0	8.6	0.65	14.6	2.55	0.80	2.6	378	122	1/5-12/11	90	60
Head Cabbage – pre Decembe 31st	60.0	8.6	0.48	10.8	2.55	0.80	2.7	288	44	18/5-19/7	90	60
Head Cabbage post December 31st	53.0	8.6	0.46	10.0	2.55	0.80	2.7	270	74	31/7-15/01	90	60
Collards Pre December 31st	20.0	8.6	0.34	5.1	3.45	0.60	4.0	203	51	16/7-24/9	45	60
Collards Post December 31st	30.0	8.6	0.38	6.8	3.45	0.60	3.8	260	41	15/9-15/01	60	60
Cauliflower (Over Winter)	–	–	–	8.1	3.45	0.60	3.7	300	85	30/7-10/03	75	60
Calabrese	16.3	10.4	0.17	10.0	1.80	3.50	2.3	226	36	27/04-25/06	90	60
Cauliflower summer	30.6	8.2	0.37	6.8	3.45	0.60	3.8	259	44	21/5-21/7	75	60
Lettuce (Crisp)	45.5	5.3	0.50	4.8	2.60	1.10	3.4	165	22	15/05-15/06	45	60
Radish	50.0	–	–	–	–	–	–	100ₖ	24	2/05-11/06	30	60
Bulb onions spring	60.5	12.7	0.81	9.4	1.20	3.50	1.6	147	20	13/03-12/05	60	60
Bulb onions overwintered	60.5	12.7	0.81	9.4	1.20	3.50	1.6	147	20	as above	60	60
Salad onions	30.0	12.7	0.81	4.7	1.20	3.50	2.4	114	20	as above	30	60
Salad onions overwintered	30.0	12.7	0.81	4.7	1.20	3.50	2.4	114	20	as above	30	60
Leeks	47.0	14.2	0.57	11.8	2.00	4.00	2.4	279	132	21/4-12/12	45	60
Beetroot	60.0	–	–	–	–	–	–	270*	65	18/5-16/08	60	60
Parsnips and (Turnips)	48.0	–	–	–	–	–	–	241*	92	30/03-27/08	90	60
Swede	84.8	11.7	0.62	16.0	1.35	1.87	1.4	222	92	30/3-27/08	90	60
Carrots	150.0	11.4	0.81	21.2	0.82	7.00	0.8	178	66	2/05-8/08	90	60

*N uptake taken from German KNS System 2007

Insufficient data to include asparagus, celery, peas and beans, sweet corn, courgettes, and bulbs

Appendix 11: Profit Maximisation and Social Cost

The grassland section of this book provides fertiliser recommendations based on what is required for different farming systems. Recommendations for all other crops given in this book are for the on-farm economic optimum rate of nitrogen.

Crop Yield and Revenue Curve with Variable Fertiliser Costs

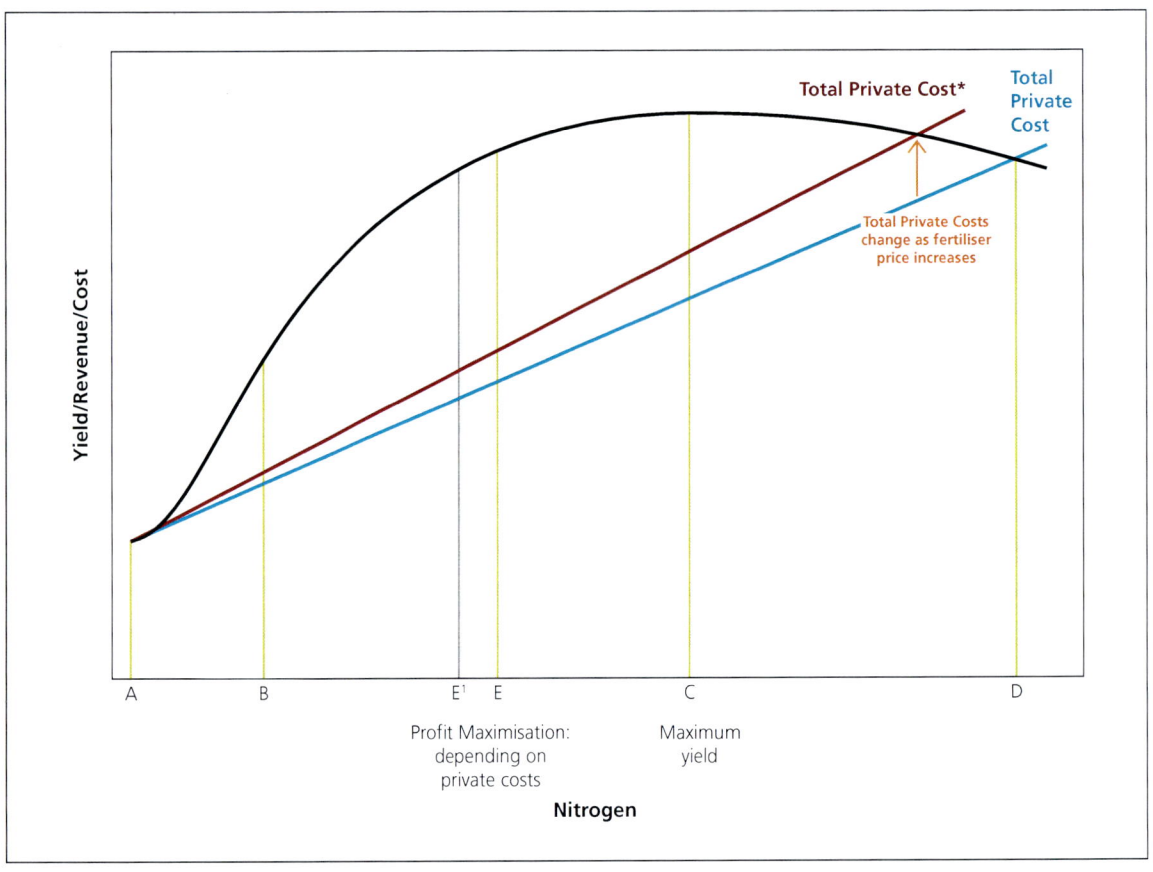

In reference to the diagram immediately above.

- Maximum profit or the on-farm economic optimum occurs between points B and C, but where precisely will vary according to the value of the crop and amongst other inputs, the cost of nitrogen.
- From point (B) to (C) the rate of increase in yields following nitrogen application is falling. In other words, there are diminishing marginal returns to fertiliser use, in terms of associated yield increases.
- The maximum yield is reached at point (C).
- After point (C) additional application of nitrogen results in no increase in yields and then falling yields which can be due to problems such as lodging of cereals etc as mentioned above.
- The yield curve can also represent the revenue provided to the farmer, if we assume that farmers are unable to affect the market price no matter how much output they produce individually.

Appendix 11: Profit Maximisation and Social Cost

- The cost of nitrogen for the farmer is shown by the total **private** cost curve. This is a linear function as the price of nitrogen is assumed to be constant (i.e. it shows costs where there are no changes to the price of nitrogen) and therefore the total cost of nitrogen increases at a constant rate as nitrogen use rises.
- So we now have a total revenue curve and a total cost curve for different levels of nitrogen application. Total profit is the difference between the total yield/revenue and total cost. The point at which these curves are furthest apart represents the maximum profit the farmer can earn and this occurs at point (E). Point (E) is therefore the on-farm economic optimum amount of nitrogen and gives the maximum profit. **This nitrogen can be supplied from fertiliser and/or organic manure.**
- At point (D) the total cost of nitrogen begins to be greater than the value of the extra crop yield produced i.e. the revenue and this makes the profit negative.
- A change in fertiliser price will cause the Total Private Cost curve to move, resulting in a different point of maximum profit. The curve (Total Private Cost*) represents an increase in the price of fertiliser with no change in the price of the crop. This will shift the profit maximising level of nitrogen usage to the left, to (E1) – the point at which cost and revenue are then furthest apart. Likewise a change in the value of the crop would cause the Total Private Revenue curve to move and this too would shift the profit maximising level.

In practice the cost of fertiliser and the value of the crop both vary over time but it would be impractical to try to provide recommendations for each scenario. A break-even ratio that accounts for the ranges of how these variables might change over time, is used therefore to calculate recommendations based on the best economic rate – for the farmer - of nitrogen. Substantial changes in the value of the crop produce or the cost of fertiliser are needed to alter the recommendations. Tables are provided nonetheless in the recommendations to help calculate the impact of changes in the price of fertiliser and of the grain sale price. Where appropriate, different recommendations are given to achieve crop quality specifications required for different markets.

Other costs of fertiliser use

The above diagram shows the benefit that fertilisers can have on crop yield. It fails however to take account of the negative impact that the application of nitrogen can have on others in society i.e. the social impact. One example is the impact that nitrates can have when leached into watercourses. Another is nitrous oxide emissions.

Whilst scientists are confident of the processes that lead to these pollutants entering the environment, the many factors that contribute to their behaviour make it difficult to quantify the true magnitude of their effect. The diagram immediately overleaf is therefore a representation of the social cost line of nitrate leached as nitrogen application increases.

- If we add together the social cost curve and the total private cost curves we get a new cost curve: the total cost, which includes the costs associated with nitrogen leaching.
- In this case the point at which total cost and total private benefit are furthest apart is the socially optimal level of nitrogen application – point E^2.

Appendix 11: Profit Maximisation and Social Cost

Crop Yield and Revenue Curve with Social Costs

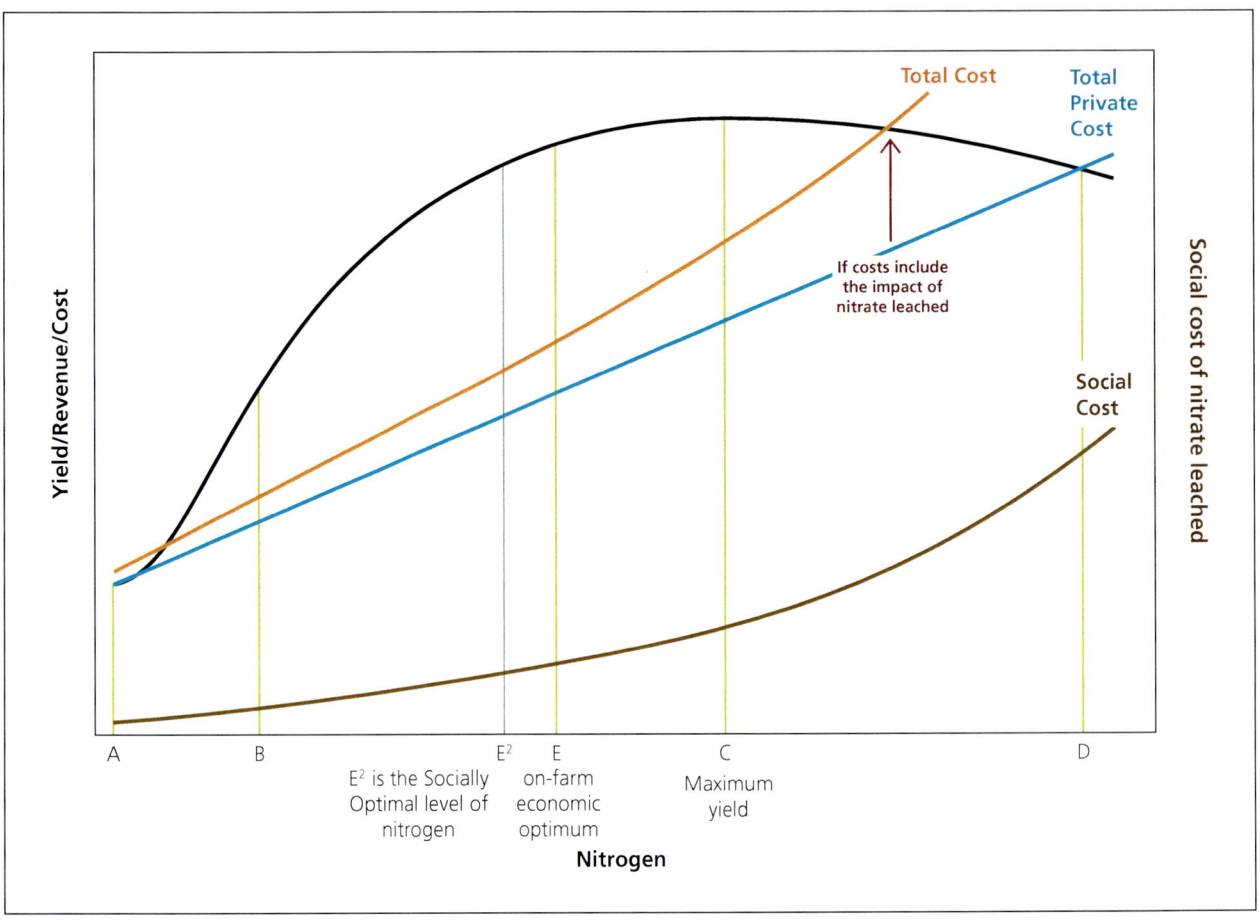

Glossary

Anion	Negatively charged form of an atom or molecule for example nitrate (NO_3^-) and sulphate (SO_4^{2-}).
Available (nutrient)	Form of a nutrient that can be taken up by a crop immediately or within a short period so acting as an effective source of that nutrient for the crop.
Bandspreading	Application of fertiliser or slurry in bands along a row of seeds or crop plants.
Biosolids	Treated sewage sludge.
Broiler/turkey litter	A mixture of bedding material and poultry excreta which is sufficiently dry to be stored in a stack without slumping.
Calcareous soil	Soil that is alkaline due to the presence of free calcium carbonate or magnesium carbonate or both.
Cation	Positively charged form of an atom or molecule for example potassium (K^+), calcium (Ca^{2+}), magnesium (Mg^{2+}) and ammonium (NH_4^+).
Cation exchange capacity	Capacity of the soil to hold cations by electrostatic forces. Cations are held at exchange sites mainly on clay particles and organic matter.
Clay	Finely divided inorganic crystalline particles in soils, less than 0.002mm in diameter.
Closed period	Period of the year when nitrogen fertilisers or certain manures should not be applied unless specifically permitted. Closed periods apply within NVZs.
Coefficient of variation (CV) (fertiliser or manure spreading)	Measure of the unevenness of application of fertilisers or manures. CV of 0% indicates perfectly even spreading, unachievable in practice. Correct operation of a well set-up spreader should give a CV of 10% for fertilisers and 25% for manures under field conditions.
Compost	Organic material produced by aerobic decomposition of biodegradable organic materials.
Content	Commonly used instead of the more accurate 'concentration' to describe nutrients in fertiliser or organic manure. For example, 6 kg N/t often is described as the nitrogen content of a manure.
Cover crop	A crop sown primarily for the purpose of taking up nitrogen from the soil and which is not harvested. Also called green manure.

Glossary

Crop available nitrogen The total nitrogen content of organic manure that is available for crop uptake in the growing season in which it is spread on land.

Crop nitrogen requirement The amount of crop available nitrogen that must be applied to achieve the economically optimum yield.

Denitrification Microbial conversion of nitrate and nitrite in anaerobic soil to nitrogen gas and some nitrous oxide.

Deposition Transfer of nutrients from the atmosphere to soil or to plant surfaces. The nutrients, mainly nitrogen and sulphur, may be dissolved in rainwater (wet deposition) or transferred in particulate or gaseous forms (dry deposition).

Digestate Organic material produced by anaerobic digestion of biodegradable organic materials. May be separated into liquid and fibre fractions after digestion.

Dirty water Lightly contaminated run-off from lightly fouled concrete yards or from the dairy/parlour that is collected separately from slurry. It does not include liquids from weeping-wall stores, strainer boxes, slurry separators or silage effluent which are rich in nitrogen and regarded as slurries.

Economic optimum (nitrogen rate) Rate of nitrogen application that achieves the greatest economic return from a crop, taking account of crop value and nitrogen cost.

Efficiency factor (manures) Percentage of total nitrogen in a manure that is available to the crop for which the manure was applied. There are mandatory minimum values in NVZs for use when estimating the nitrogen availability of manures.

Erosion Movement (transport) of the soil by running water or wind.

Eutrophication Enrichment of ecosystems by nitrogen or phosphorus. In water it causes algae and higher forms of plant life to grow too fast. This disturbs the balance of organisms present in the water and the quality of the water concerned. On land, it can stimulate the growth of certain plants which then become dominant so that natural diversity is lost.

Excess rainfall Rainfall between the time when the soil profile becomes fully wetted in the autumn (field capacity) and the end of drainage in the spring, less evapo-transpiration during this period (i.e., water lost through the growing crop).

Glossary

FACTS	UK national certification scheme for advisers on crop nutrition and nutrient management. Membership renewable annually. A FACTS Qualified Adviser has a certificate and is a member either of the FACTS Annual Scheme or of the BASIS Professional Register.
Farmyard manure (FYM)	Livestock excreta that is mixed with straw bedding material that can be stacked in a heap without slumping.
Fertiliser	See **Manufactured fertiliser.**
Fluid fertiliser	Pumpable fertiliser in which nutrients are dissolved in water (solutions) or held partly as very finely divided particles in suspension (suspensions).
Frozen hard	Soil that is frozen for more than 12 hours. Days when soil is frozen overnight but thaws out during the day do not count.
Granular fertiliser	Fertiliser in which particles are formed by rolling a mixture of liquid and dry components in a drum or pan. Typically, particles are in the 2 – 4mm diameter range.
Grassland	Land on which the vegetation consists predominantly of grass species.
Greenhouse gas	Gas such as carbon dioxide, methane or nitrous oxide that contributes to global warming by absorbing infra-red radiation that otherwise would escape to space.
Green manure	See **Cover crop.**
Heavy metal	Cadmium, copper, lead, mercury, nickel or zinc. Elements that are potentially toxic to mammals above critical levels. Copper, nickel and zinc are required by plants in very small amounts.
Incorporation	A technique (discing, rotovating, ploughing or other methods of cultivation) that achieves some mixing between an **organic manure** and the soil. Helps to minimise loss of nitrogen to the air through **volatilisation**, and nutrient **run-off** to surface waters.
Inorganic fertiliser	**Manufactured fertiliser** that contains only inorganic ingredients or urea.
Layer manure	Poultry excreta with little or no bedding.
Leaching	Process by which soluble materials such as nitrate or sulphate are removed from the soil by drainage water passing through it.

Glossary

Ley
Temporary grass, usually ploughed up one to five years (sometimes longer) after sowing.

Lime requirement
Amount of standard limestone needed in tonnes/ha to increase soil pH from the measured value to a higher specified value (often 6.5 for arable crops). Can be determined by a laboratory test or inferred from soil pH.

Liquid fertiliser
See **Fluid fertiliser.**

Livestock manure
Dung and urine excreted by livestock or a mixture of litter, dung and urine excreted by livestock, even in processed organic form. Includes FYM, slurry, poultry litter, poultry manure, separated manures, granular or pelletised manures.

Macronutrient
See **Major nutrient** or **Secondary nutrient**.

Maintenance application (phosphate or potash)
Amount of phosphate or potash that must be applied to replace the amount removed from a field at harvest (including that in any straw, tops or haulm removed).

Major nutrient
Nitrogen, phosphorus and potassium that are needed in relatively large amounts by crops (see also Secondary nutrients and Micronutrients).

Manufactured fertiliser
Any fertiliser that is manufactured by an industrial process. Includes conventional straight and NPK products (solid or fluid), organo-mineral fertilisers, rock phosphates, slags, ashed poultry manure, liming materials that contain nutrients.

Manure
See **Livestock manure** and **Organic manure.**

Micronutrient
Boron, Copper, Iron, Manganese, Molybdenum, Zinc that are needed in very small amounts by crops (see also **Major nutrients** and **Micronutrients**). Cobalt and selenium are taken up in small amounts by crops and are needed in human and livestock diets.

Mineral nitrogen
Nitrogen in ammonium and nitrate forms.

Mineralisable nitrogen
Organic nitrogen that is readily converted to ammonium and nitrate by microbes in the soil, for example during spring.

Mineralisation
Microbial breakdown of organic matter in the soil, releasing nutrients in crop-available, inorganic forms.

Neutralizing value (NV)
Percentage calcium oxide (CaO) equivalent in a material. 100kg of a material with a neutralising value of 52% will have the same neutralising value as 52kg of pure CaO. NV is determined by a laboratory test.

Glossary

Nitrogen uptake efficiency — Uptake of nitrogen from soil, fertiliser or manure expressed as a percentage of nitrogen supply from that source.

Nitrogen use efficiency — Ratio of additional yield produced to the amount of nitrogen applied to achieve that increase. Often expressed as kg additional yield per kg N applied.

Nitrous oxide (N_2O) — A potent greenhouse gas that is emitted naturally from soils. The amount emitted is related to supply of mineral nitrogen in the soil so increases with application of manures and fertilisers, incorporation of crop residues and growth of legumes and is greater in organic and peaty soils than in other soils.

Offtake — Amount of a nutrient contained in the harvested crop (including straw, tops or haulm) and removed from the field. Usually applied to phosphate and potash.

Olsen P — Concentration of available P in soil determined by a standard method (developed by Olsen) involving extraction with sodium bicarbonate solution at pH 8.5. The main method used in the England, Wales and Northern Ireland and the basis for the **Soil Index** for P.

Organic manure — Any bulky organic nitrogen source of livestock, human or plant origin, including livestock manures.

Organic soil — Soil containing between 10% and 20% organic matter (in this *Manual*). Elsewhere, sometimes refers to soils with between 6% and 20% organic matter.

Peaty soil (peat) — Soil containing more than 20% organic matter.

Placement — Application of fertiliser to a zone of the soil usually close to the seed or tuber.

Poultry litter — See **Broiler/turkey litter**.

Poultry manure — Excreta produced by poultry, including bedding material that is mixed with excreta, but excluding duck manure with a readily available nitrogen content of 30% or less.

Prilled fertiliser — Fertiliser in which particles (prills) are formed by allowing molten material to fall as droplets in a tower. Droplets solidify during the fall and tend to be more spherical and somewhat smaller than granules (see **Granular fertiliser**).

Glossary

Readily available nitrogen	Nitrogen that is present in livestock and other organic manures in molecular forms that can be taken up immediately by the crop (ammonium or nitrate, or in poultry manure uric-acid N). High in slurries and poultry manures (typically 35-70% of total N) and low in FYM (typically 10-25%).
Removal	See **Offtake**.
Run-off	Movement of water across the soil surface which may carry nutrients from applied manures or fertilisers and with soil particles.
Safe Sludge Matrix	Guidance on use of sewage sludge for different crops agreed by Water UK and the British Retail Consortium.
Sand	Soil mineral particles larger than 0.05mm.
Secondary nutrient	Magnesium, sulphur, calcium or sodium that are needed in moderate amounts by crops.
Silt	Soil mineral particles in the 0.002 – 0.05mm diameter range.
Slurry	Excreta of livestock (other than poultry), including any bedding, rainwater and washings mixed with it, that can be pumped or discharged by gravity. The liquid fraction of separated slurry is also defined as slurry.
SNS Index	**Soil Nitrogen Supply** expressed in seven bands or Indices, each associated with a range in kg N/ha.
Soil Index (P, K or Mg)	Concentration of available P, K or Mg, as determined by standard analytical methods, expressed in bands or Indices.
Soil Mineral Nitrogen (SMN)	Ammonium and nitrate nitrogen measured by the standard analytical method and expressed in kg N/ha.
Soil Nitrogen Supply (SNS)	The amount of nitrogen (kg N/ha) in the soil that becomes available for uptake by the crop in the growing season, taking account of nitrogen losses.
Soil organic matter	Often referred to as humus. Composed of organic compounds ranging from undecomposed plant and animal tissues to fairly stable brown or black material with no trace of the anatomical structure of the material from which it was derived.
Soil texture	Description based on the proportions of sand, silt and clay in the soil.

Glossary

Soil type	Description based on **soil texture**, depth, chalk content and organic matter content.
Solid manure	Organic manure which can be stacked in a freestanding heap without slumping.
Target Soil Index	Lowest soil P or K Index at which there is a high probability crop yield will not be limited by phosphorus or potassium supply. See **Soil Index (P, K or Mg)**.
Tillage land	Land that is not being used for grass production and is sown with a crop.
Trace element	See **Micronutrient**.
Volatilisation	Loss of nitrogen as ammonia from the soil to the atmosphere.
Water-soluble phosphate	Phosphate, expressed as P_2O_5, that is measured by the statutory method for fertiliser analysis. Not necessarily a measure of available phosphate – high water-solubility indicates high availability but low water-solubility does not necessarily indicate low availability.
Weathering	Breakdown of soil mineral particles by physical or chemical processes. Enhanced by variation in temperature and moisture. A significant mechanism for release of potassium from clay minerals.

Notes

Notes

Notes